Polarized Proton Ion Sources
(Ann Arbor, 1981)

AIP Conference Proceedings
Series Editor: Hugh C. Wolfe
Number 80

Polarized Proton Ion Sources
(Ann Arbor, 1981)

Editors
A. D. Krisch
A. T. M. Lin
University of Michigan

American Institute of Physics
New York 1982

L.C. Catalog Card No. 82-71025
ISBN 0-88318-179-7
DOE CONF- 8105145

TABLE OF CONTENTS

Introduction

A.D. Krisch
Randall Laboratory of Physics
University of Michigan

A workshop on High Intensity Polarized Proton Ion Sources was held at The University of Michigan at Ann Arbor on May 20-27, 1981. The increased interest in intermediate and high energy polarized proton beams made such a workshop seem especially appropriate at this time.

The workshop tried to foster new ideas for polarized proton ion sources, evaluate existing ideas and perhaps dispose of some unworkable concepts. We attempted to reach some conclusions on how to produce the best reliable source for ongoing projects and some conclusions about very high intensity sources for the future.

A number of topics were discussed including:

- Cold Atomic Hydrogen Sources
- Optical Pumping Sources (Dye-Lasers)
- Colliding Beam Charge Exchange Sources
  - High Intensity Cs° Beams
  - High Intensity H^- and D^- Beams
  - Improved Atomic Beam Stage
- Lamb Shift Sources

Another purpose of this workshop was to bring the polarized ion source experts together with some accelerator experts. This is especially appropriate because the large and expensive high energy accelerators put very strong reliability demands on sources; and we feel that both the source and accelerator experts benefited by close scientific and personal contact for 8 days. We were especially pleased that a few experts on promising new areas also attended to inform the workshop of their ideas.

The workshop consisted of a few short lectures by experts each day followed by small working groups of typically 12 people. A total of about 30 people participated and most attended for the entire 8 days. The basic idea of this workshop was to get a small group of the world's experts on each relevant subject to work closely together, in a concentrated and intense way, for 7 working days. It was a somewhat exhausting experience but we think that it produced some very important scientific results.

The organizing committee for this workshop was W. Haeberli, Wisconsin; V.W. Hughes, Yale; A.D. Krisch, Michigan (Chairman); and D.R. Moffett, Argonne. The workshop was sponsored by the International Committee for Symposia on High Energy Spin Physics:

A.D. Krisch-Michigan (Chairman)
A. Abragam-Coll. de France
O. Chamberlain-Berkeley
E.D. Courant-Brookhaven
G. Fidecaro-CERN
V.W. Hughes-Yale
M. Jacob-CERN
A. Maisaike-KEK
L. Michel-Bures-sur-Yvette
A.N. Skrinsky-Novosibirsk
L.D. Soloviev-Serpukhov
G.A. Voss-DESY

The workshop was an excellent example of the benefits which can come from direct cross-fertilization between different scientific disciplines. D. Kleppner and other Atomic Physicists brought to the accelerator people an understanding of the recent breakthroughs in spin polarized cold atomic hydrogen. The Nuclear and High Energy people fed back to the Atomic people their needs and their scientific and technical capabilities. There is now some hope that this intense interaction has suggested a scheme for a polarized ion source much more intense than those presently available. Such a source would give a polarized proton beam of intensity equal to unpolarized beams. This would allow proton accelerators and storage rings to run polarized all the time, with experimenters uninterested in spin simply averaging over the different spin states. We do not yet have all the hardware to attain this happy state of affairs, but most participants left the workshop feeling considerably closer to this goal.

List and Addresses of Participants at the
Workshop on High Intensity Polarized Proton Ion Sources
Ann Arbor, Michigan, May 20-27, 1981

1. L.W. Anderson
Dept. of Physics
University of Wisconsin
1150 University Ave
Madison, WI 53706
(608) 262-8962

2. J. Arvieux
Laboratoire National Saturne
91191 Gif-sur-Yvette
FRANCE
(6)908-3121 or
908-2203

3. E.P. Chamberlin
M.S. 823,
Los Alamos National Lab.
P.O. Box 1663
Los Alamos, NM 87545
(505)667-4593

4. G. Clausnitzer
University of Giessen
6300 Giessen
W. Germany
(0641)702-2655

5. T.B. Clegg
Dept. of Physics & Astronomy
Univ. of N.C.
Chapel Hill, NC 27514
(919) 933-3016

6. W. Cornelius
MP-10 MS 841
Los Alamos National Lab
Los Alamos, NM 87545
(505) 667-5720

7. M.A. Cummings*
Dept. of Physics
University of Michigan
Ann Arbor, MI 48109
(313)764-4443

8. L. Dick
CERN - EP Division
CH-1211 Geneva 23
Switzerland
(022) 832-2261

9. T.R. Donoghue
Dept. of Physics
Ohio State University
1302 Kinnear Rd.
Columbus, OH 43212
(614)422-9524

10. W. Gruebler
Laboratorium fur Kernphysik
Eidg. Technische Hochschule
Honggerberg-CH-8093 Zurich
Switzerland
(01) 377-4411

11. W. Haeberli
Dept. of Physics
University of Wisconsin
1150 University Ave.
Madison, WI 53706
(608) 262-0009

12. E. Hinds
Sloane Physics Lab
217 Prospect St.
New Haven, CT 06520
(203) 436-8671

13. V.W. Hughes
Dept. of Physics
Yale University
P.O. Box 6666
New Haven, CT 06511
(203)436-3566

14. S. Jaccard
SIN
CH-5234, Villigen
Switzerland
(056) 99 31 11

15. D. Kleppner
Department of Physics
26-231, M.I.T.
Cambridge, MA 02139
(617) 253-6811

16. A. Kponou
Building 911B
Brookhaven National Lab.
Upton, L.I., NY 11973
(516) 345-7254

17. A.D. Krisch
Randall Lab. of Physics
University of Michigan
Ann Arbor, MI 48109
(313) 764-4443

18. W. Kubischta
CERN- E.P. Div.
CH-1211, Geneva 23
Switzerland
(022) 83 25 53

19. Y.Y. Lee
Accelerator Dept.
Bldg. 911B
Brookhaven National Lab.
Upton, NY 11973
(516)345-4663

20. L. Levy*†
Randall Lab. of Physics
University of Michigan
Ann Arbor, MI 48109

21. A.M.T. Lin
Randall Lab. of Physics
University of Michigan
Ann Arbor, MI 48109
(313) 764-4443

22. S.R. Magill*
Building 362
Argonne National Lab
Argonne, IL 60439

23. D. Lowenstein
Accelerator Dept.
Brookhaven National Lab.
Upton, L.I., NY 11973
(516)345-4611

24. R. Mobley
Building 911B
Brookhaven National Lab.
Upton, L.I., NY 11973
(516) 345-7254

25. D.R. Moffett#
D362
Argonne National Laboratory
Argonne, IL 60439

26. Y. Mori
KEK, Nat'l. Lab. for
High Energy Phys.
Oho-Machi, Tsukuba-gun
Ibaraki-ken, JAPAN
298-64-1171

27. D.E. Murnick
Bell Laboratories
Murray Hill, NJ 07974
(201) 582-4825

28. D.C. Peaslee‡
Division of High Energy Phys
U.S. Department of Energy
Washington, DC 20545
FTS: 223-3367

29. P. Quin
Department of Physics
University of Wisconsin
Madison, WI 53706
(608) 262-8739

30. R. Raymond
Nuclear Physics Lab
Univ. of Colorado
Campus Box 446
Boulder, CO 80309
(303) 492-7483

31. P. Schiemenz
Sek. Phys. der
Universitat Munchen
Am Coulomb wall 1
8046 Garching, W. Germany
(089) 3209 5130

32. P.W. Schmor
TRIUMF, U.B.C.
4004 Wesbrook Mall
Vancouver, B.C.
Canada V6T 2A3
(604) 228-4711

33. P.F. Schultz#
Building 360
Argonne National Lab.
Argonne, IL 60439
(312) 972-6255

34. Th. Sluyters
Accelerator Dept. Bldg. 911
Brookhaven National Lab.
Upton, L.I., NY 11973
(516) 345-4785

35. E. Steffens
MPI fur Kernphysik
Postf. 103980
D-69 Heidelberg
W. Germany
(06221) 516 408

36. R.L. York
Los Alamos National Lab.
M.S. 823, P.O. Box 1663
Los Alamos, NM 87545
(505) 667-4577

Secretaries

Anne Allen
Alice Carroll

---

*Scientific Secretaries
†present address: Physics Dept., Cornell University
#present address: Bell Labs, Naperville, IL
‡present address: Physics Dept., University of Maryland

PROGRAM

Workshop on

HIGH INTENSITY POLARIZED PROTON ION SOURCES

UNIVERSITY OF MICHIGAN, ANN ARBOR

May 20 - 27 1981

Meeting Rooms: 1035 and 1041 Randall Lab of Physics

UPDATE May 25, 1981

Wednesday May 20, 1981

9:00 Welcome and Introduction (1041) — A.D. Krisch
9:30 Review of Lamb Shift PPIS (1041) — T.B. Clegg
10:45 Coffee
11:15 Review of Atomic Beam Stage of Ground State PPIS — W. Gruebler
12:30 Lunch
2:00 Review of Ionization Stage of Ground State PPIS — W. Haberli
3:15 Tea
3:45 Improved Atomic Beam Stage Working Group (1035) W. Gruebler (McGill) | Lamb Shift PPIS Working Group (1041) T.B. Clegg (Cummings)
5:15 End Sessions
7:00 Reception Krisch's house 3616 Chatham Way

Thursday May 21, 1981

9:15 Review of Possible Uses of Cold Atomic Hydrogen in PPIS — D. Kleppner
10:30 Coffee
11:00 Cold Atomic Hydrogen Working Group (1035) D. Kleppner (Levy) | Ionization Stage Working Group (1041) W. Haberli (McGill)
12:30 Lunch
2:00 Review of High Intensity Unpolarized Cs°, H^-, and D^- Sources (1041) — T. Sluyters
3:15 Tea
3:45 High Intensity Unpolarized Source Working Group (1035) T. Sluyters (McGill) | Lamb Shift PPIS Working Group (1041) T.B. Clegg (Cummings)
5:15 End Sessions

Friday May 22, 1981

9:15 Review of Optical Pumping PPIS (1041) — L.W. Anderson
10:30 Coffee
11:00 Optical Pumping PPIS Working Group (1041) L.W. Anderson (Levy) | High Intensity Unpolarized Source Working Group (1041) T. Sluyters (McGill)
12:30 Lunch
2:00 Cold Atomic Hydrogen Working Group (1035) D. Kleppner (Levy)
3:30 Tea
4:00 Optical Pumping PPIS Working Group (1035) L.W. Anderson (Cummings) | Ionization Stage Working Group (1041) W. Haberli (McGill)
5:00 End Sessions
8:00 Play "Once Upon a Mattress"

Saturday May 23, 1981

9:45 Roundtable Discussion (1041) — A.D. Krisch
11:00 Leave for Greenfield Village Tour and Lunch (from 1041)
4:00 End Tour

Monday May 25, 1981

9:15 Progress on Lamb-Shift PPIS (1041) — T.B. Clegg
10:30 Coffee
11:00 Lamb-Shift PPIS Working Group (1035) T.B. Clegg (Cummings) | $H^+ \rightarrow H^-$ Ionization Working Group (1041) W. Gruebler (McGill)
12:30 Lunch
2:00 Polarized Protons Saclay — J. Arvieux
2:30 Roundtable Discussion — A.D. Krisch
3:30 Tea
4:00 Improved Atomic Beam Stage Working Group (1041) W. Gruebler (McGill) | Optical Pumping PPIS Schmor (1035) (Cummings)
5:30 End Sessions

Tuesday May 26, 1981

9:15 Progress on Improved Atomic Beam Stage (1041) — W. Gruebler
10:30 Coffee
11:00 Ionization Stage Working Group (1041) S. Jaccard (McGill) | Lamb Shift PPIS Working Group (1035) P. Schiemenz (Cummings)
12:30 Lunch
2:00 Progress on Ionization Stage — S. Jaccard
3:15 Tea
3:45 Improved Atomic Beam Stage Working Group (1041) L. Dick (McGill) | Cold Atomic Hydrogen Working Group (1035) D. Kleppner (tentative) (Levy)
5:15 End Sessions
7:00 Chinese Banquet Hung-Wan Restuarant

Wednesday May 27, 1981

9:45 Polarized Gas Jets (1041) — W. Kubishta
10:15 Polarized Protons at KEK (1041) — Y. Mori
10:45 Coffee
11:15 Roundtable Discussion (1041) — A.D. Krisch
12:30 Lunch
2:00 Workshop Summary (1041) — V.W. Hughes
3:15 End Session

SUMMARY TALK

Vernon W. Hughes
Gibbs Laboratory, Yale University, New Haven, CT. 06511

1. INTRODUCTION

I had hoped that my role in our Workshop Program would be to give the afterdinner speech, but I was unable to attend the banquet and in any case was only invited by Alan Krisch to give the Summary Talk. Alan is so conscientious - almost to a fault - in his conduct of a workshop that one certainly must do what he asks with good spirit.

Our workshop on High Intensity Polarized Proton Ion Sources has concentrated almost exclusively on the methods and technologies of proton ion sources. As part of the record of this workshop it seems to me some brief remarks on the physics results and motivations associated with polarized beams from accelerators are appropriate. Although we have concentrated principally on polarized proton (and deuteron) beams, other particles - electrons, muons, and heavier ions - are of course also of interest.

A short list of general topics in particle and nuclear physics to which the use of accelerated polarized beams has contributed decisive, unique, and important information is given in Table I. References to the very large body of work in polarization phenomena in particle and nuclear physics are conveniently given to two recent major conferences - the 1980 Lausanne Conference[1] on high energy particle physics and the 1980 Sante Fe Conference[2] on nuclear physics.

Table I Some Topics in Polarization Phenomena

Particle Physics
- (1) Invariance principles
  - Parity violation in e-d and p-p scattering
  - T invariance tests ($e^- + p_\uparrow \rightarrow e^- + X$)
- (2) Nucleon spin structure
  - Polarized e-p scattering
- (3) Spin dependent strong interactions
  - Polarized p-p elastic scattering with high momentum transfer
  - Dibaryon
- (4) Polarized protons in a high energy storage ring
  - Disentangle strong and weak interactions

Nuclear Physics
- (5) Nucleon-nucleon interaction
  - Polarized p-p and n-p scattering
- (6) Nuclear structure and reactions
  - Polarized p-nucleus (optical potential)
  - Nuclear spectroscopy
  - Nuclear reactions (phase shift analyses)

The reason for this Workshop is our firm belief that polarized beams can be expected to contribute new information in the future as well. Indeed as particle physics and nuclear physics become more advanced, we anticipate that polarization phenomena will become relatively more important. One example in particle physics with ISABELLE[3] is the use of the spin dependence of a reaction to distinguish weak from electromagnetic and strong interactions. Hence we have good reason to work hard to improve the quality of our sources of polarized particles.

The first international conference on what we now call particle and nuclear spin physics was the International Symposium on Polarization Phenomena of Nucleons held at Basel, Switzerland in 1960.[4] A partial Table of Contents for this Symposium is given in Table II.

Table II Table of Contents (Partial)

Proceedings of the International Symposium on Polarization Phenomena of Nucleons, Basel, July 4-8, 1960, Editors P. Huber and K.P. Meyer, Helv. Phys. Acta, Supplementum VI (Birkhäuser Verlag, Basel, 1961).

....

IV. Theories Concerning Polarization Effects of Nucleons

Section I on Production of Sources and Targets of Polarized Nuclei includes discussion of most of the principles of our present-day polarized proton and deuteron sources. At the time of that Symposium several groups were reporting their first success in accelerating polarized protons or deuterons to energies of about 100 keV or higher. Much of the theoretical interest at that Symposium, as indicated under Section IV of the Table of Contents, was on polarization effects in the nucleon-nucleon interaction and in nuclear scattering at rather low energies. It is interesting to note that a number of people at our Workshop (G. Clausnitzer, L. Dick, W. Haeberli, and V.W. Hughes) were participants at the Basel Symposium.

Fortunately much of the basic information, including considerable technical detail, on polarized proton and deuteron sources now in operation with accelerators or under development has been published recently in the Lausanne[1] and Santa Fe[2] Conference Proceedings. Relevant titles on polarized sources are listed in Tables III and IV.

Although our Workshop did not include polarized electron sources, some results and problems in this field are relevant to polarized proton sources. In section 4 brief comments and references on polarized electron sources are given.

This summary talk includes the following sections; 2. Status of methods used for operational polarized proton sources, 3. Other approaches to polarized proton sources, including the use of low temperature techniques, 4. Polarized electron sources, 5. New accelerator-related projects, and 6. Conclusions.

Table III References on Polarized Sources from 1980 Lausanne Conference

<u>High-Energy Physics with Polarized Beams and Polarized Targets</u>, Lausanne Symposium, 1980, Editors C. Joseph and J. Soffer, Experentia Supplementum Vol. 38 (Birkhäuser Verlag, Basel, 1981).

		Page
T.O. Niinikoski	Progress in Polarized Targets	191
W. Haeberli	Sources of Polarized negative ions: Progress and Prospects	199
W. Kubischta	The CERN Polarized Atomic Hydrogen Beam Target	212
Yu K.Pilipenko	Cryogenic Source and Ionizer for a Beam of Polarized Deuterons	429
W. Grüebler	A High-intensity Polarized Ion Source for Negative Hydrogen Ions	432
D.R. Moffett	Polarized Negative Hydrogen Source for the AGS	435
Y. Mori	Optically Pumped Na Atoms for Intense Polarized H^- Ion Source	439
S. Jaccard	Beam Polarization Tuning at Injection Energy	443
K.P. Schüler	A Proposed New Technique for Polarized Electron-Polarized Nucleon Scattering	460

Table IV References on Polarized Sources from 1980 Santa Fe Conference

Polarization Phenomena in Nuclear Physics - 1980, (Fifth International Symposium, Santa Fe), Editors G.G. Ohlsen, Ronald E. Brown, Nelson Jarmie, W.W. McNaughton and G.M. Hale, AIP Conference Proceedings No. 69, Part 2 (American Institute of Physics, New York, 1981).

	Page
Polarized Electron Sources G. Baum	785
New Technique for the Production of Polarized ^{3}He Ions R.J. Slobodrian	797
Lasers in the Production of Polarized Beams and Targets D.E. Murnick and M.S. Feld	804
Absolute Polarization Standards at Medium and High Energies M.W. McNaughton	818
Survey of Methods for Rapid Spin Reversal J.L. McKibben	830
New Developments in Polarized Ion Sources W. Grüebler and P.A. Schmelzbach	848
Rapporteur's Report: Polarized Beam and Target Technology, Techniques, and Formalism P.A. Quin	866
The Wisconsin Colliding-Beam Negative Polarized-Ion Source W. Haeberli, M.D. Barker, G. Caskey, C.A. Gossett, D.G. Mavis, P.A. Quin, J. Sowinski, and T. Wise	877
New Type Polarized H^- Ion Source for the KEK 12 GeV Synchrotron Y. Mori, K. Ito, A. Takagi, C. Kubota, and S. Fukumoto	879
The Duoplasmatron Emission Aperture: A Very Critical Parameter in Lamb-Shift Sources W. Arnold, H. Berg, and G. Clausnitzer	882
An Upgrading Program for the TUNL Lamb-Shift Polarized Source T.B. Clegg, S.M. Mitchell, R.L. Varner, W.R. Wylie, S.A. Wender, C.E. Floyd, R.K. Murphy, and M.E. Wright	884
Improvements in the LAMPF Lamb-Shift Polarized Source E.P. Chamberlin, R.L. York, H.E. Williams, and E.L. Rios	887
The University of Manitoba Polarized D^- and H^- Source S. Oh, M. De Jong, J. Bruckshaw, I. Gusdal F. Konopasek, A. McIlwain, and R. Pogson	890
The Munich Polarized Ion Source P. Schiemenz, D. Ehrlich, R. Frick, G. Graw, and U. Meyer-Berkhout	893
Operating Experience and Cesium Recycling on the LASL Polarized Triton Source R. Hardekopf	896

2. STATUS OF METHODS USED FOR OPERATIONAL POLARIZED PROTON SOURCES

Current operational polarized proton sources are based either on the use of ground 1S state H atoms (Fig. 1) or of metastable 2S state H atoms (Fig. 2). In this volume the ground state sources are reviewed by W. Grüebler, W. Haeberli, and S. Jaccard, and the metastable state sources by T.B. Clegg.

Typical operating conditions for ground state sources are given in Table V for production of both polarized H^+ and H^- beams. Examples of operational sources are the polarized H^+ source for the SIN accelerator in Villigen, Switzerland, the polarized H^- source at the ETH tandem electrostatic accelerator in Zurich which employs ionization of H by e^- followed by double e^- capture by fast H^+ in alkali vapor, and the polarized H^- source at the Wisconsin tandem electrostatic accelerator which employs collisions of H with fast Cs atoms to produce H^-.

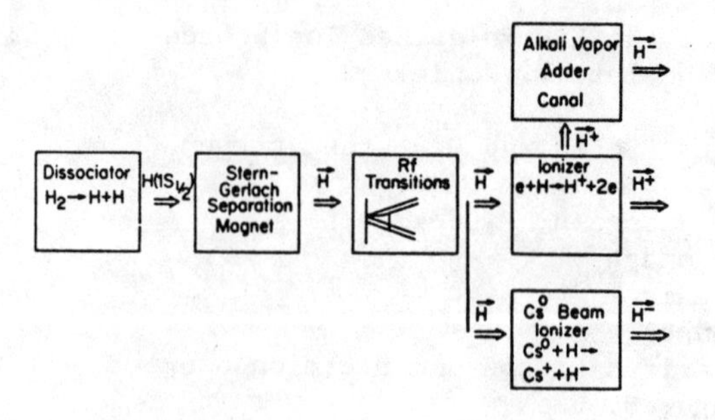

Fig. 1. Schematic diagram of the atomic beam sources for polarized hydrogen ions.

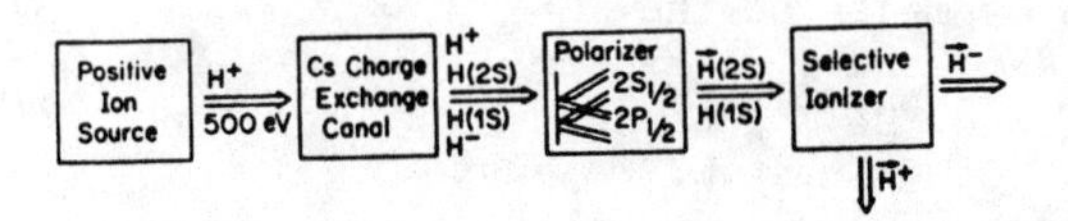

Fig. 2. Schematic diagram of the Lamb-shift polarized ion source for hydrogen ions.

Table V Characteristic Operating Conditions for Polarized Proton Sources

Ground State Sources
- H^+
  - Intensity, I: 100 μA
  - Polarization, P: 80%
  - Emittance: 10π mm-mrad $(MeV)^{1/2}$
- H^-
  - I = 3 μA
  - P = 80%
  - Emittance: 5π mm-mrad $(MeV)^{1/2}$

Metastable State Sources
- H^-
  - I = 1 μA
  - P = 70%
  - Emittance = 3π mm-mrad $(MeV)^{1/2}$

Typical operating conditions for metastable state sources are also given in Table V. Examples are the polarized H^- source at the TUNL tandem electrostatic accelerator and that at the LAMPF 800 MeV proton linear accelerator.

For the ground state sources discussions about improvements emphasized possible increase in the atomic beam density with lower temperature sources and increase in the ionization efficiencies.

Predicting the characteristics of the atomic beam from the source is an old and difficult problem because the source usually operates in the transition region between molecular and viscous flow.[5,6,7,8] Furthermore, we are interested not only in the total intensity (atoms per sec) from the orifice but also in the angular distribution of the beam and in its molecular content. Basic to an understanding of

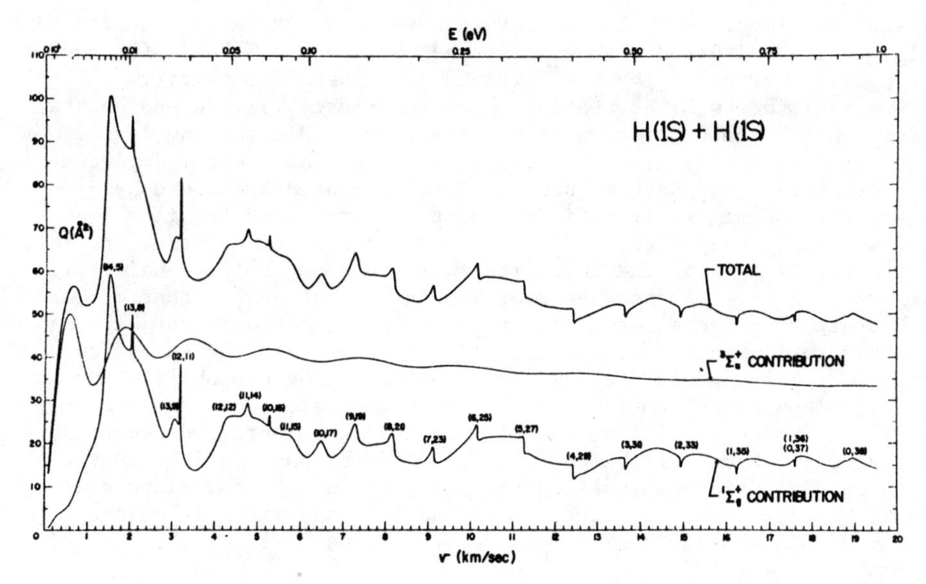

Fig. 3. Dependence of the total elastic scattering cross section Q(A°^{2}), and the separate singlet and triplet contributions to Q, upon the relative velocity v(km/sec). Upper scale: collision energy in eV. The orbiting resonances are labeled by the quantum numbers (v,J) of the corresponding quasibound states.

source operation are the cross sections[9] for H - H, H - H_2, and H_2 - H_2 scattering at the relevant energies in the source. Calculated values[10] for the total elastic scattering cross section in H - H collisions are shown in Fig. 3. Because of the interest in using cold or ultracold beams, the predicted drop in the cross section at energies below 0.005 eV is important. The differential cross section as well as the total cross section is relevant to the beam characteristics. The use of multicapillary sources at low temperatures may be worth pursuing.

Ionization of the ground state H beam can be done by either of the methods indicated in Fig. 1. Conservative estimates of possible improvements in the ionization efficiencies do not suggest gains larger than about 2 by these methods. However, use of the ionizing collision $H + D^- \rightarrow H^- + D$ is attractive[11] because of its large cross section, and several schemes were considered promising for avoiding the serious problem of space charge blow-up.

A major improvement in the intensity of the polarized H^+ beam from the ground state sources does appear to be possible now with the standard method shown in Fig. 1, provided the source is cooled to LHe temperature (T = 4K). Frequent attempts have been made to cool the source to LN_2 temperature with moderate increase in the intensity of the H^+ beam, and at ANL cooling of the Pyrex nozzle of the discharge tube by contact with a copper block at 30K has proved helpful. Recent success in the measurement[12] of the hfs transition of gaseous atomic hydrogen at 4.2K, which involved transport of H from an rf dissociator immersed in LN_2 through a Pyrex tube at 4K to a quartz bulb, indicates that a beam of H can be obtained at $T = 4K$. (Molecular hydrogen is cryopumped by the Pyrex tube.) The H intensity, I, (atoms/sec), that will be available for the source will depend on the atomic and molecular cross sections and flow conditions and is not known at present. If the scattering cross sections were independent of energy, then the Knudsen number (mean free path λ divided by aperture diameter) would be independent of energy and the flow condition would be independent of T, provided the density of atoms, n, in the source is fixed. The intensity would then be proportional to $\sqrt{T}$. (molecular flow). We note however in Fig. 3 that σ_{tot} (elastic) for H - H scattering decreases at lower energies and hence it may be possible to increase n at lower T and still maintain the same flow condition, which would increase I. The ionization probability for an atom is proportional to $1/\sqrt{T}$, and hence compensates for the decrease in I. The principal increase will come from the increased acceptance of the six-pole magnet system, which is proportional to 1/T. Our workshop discussions emphasized that the design of a polarized source should be considered in its entirety, so that the six-pole magnet system should of course be matched to the temperature of the atomic H beam.

At present the metastable state H^- sources operate with somewhat lower currents and polarizations than have been achieved with the ground state sources. The fundamental limitation on the intensity of the H(2S) beam due to quenching by electric fields in the beam is not well understood. Although use of an electron cyclotron resonance H^+ beam source may be promising, a clear potential for substantial improvement in polarized H^- intensity does not appear available.

3. OTHER APPROACHES TO POLARIZED PROTON SOURCES

Two newer approaches to polarized proton sources were discussed extensively: (1) The use of cold hydrogen techniques (reviewed by D. Kleppner in this volume) and (2) Optical pumping. (reviewed by L.W. Anderson in this volume) Both of these approaches are in a rather early stage of conceptual design and development.

The ideas and promise associated with the use of cold hydrogen techniques provided the most exciting and stimulating topic at the workshop. The recent, spectacular development of these techniques has been stimulated by the aim of obtaining a Bose-Einstein gas of electron spin-polarized H atoms. One important application of cold hydrogen techniques is the use of a source of H atoms at LHe temperature (~4K) in a conventional polarized atomic beam system. (See section 2, and paper by D. Kleppner in this volume) More radical and dramatic possibilities exist of using the high density of proton spin-and electron spin-polarized H atoms achieved in a very low temperature-high magnetic field regime (0.3K, 10T) to produce an intense beam of proton spin-polarized H atoms (perhaps monochromatic), or as a polarized proton target.[13]

The optical pumping method of producing polarized H^- involves producing a gas target of electron spin-polarized Na atoms by use of a CW dye laser. Unpolarized H^+ with an energy of 5 keV capture an electron from Na, pass diabatically through zero magnetic field to produce proton spin-polarized H, and then capture a second electron to produce H^-. Although this optical pumping approach may have the potential of high H^- current, there appear to be substantial problems, including inadequate laser bandwidth to saturate the Doppler-broadened Na line, depolarization due to electron capture into excited states of H, and the large emittance of the H^- beam due to the large magnetic field to be used in the region where H^+ capture an electron from Na. At KEK where a polarized H^- beam is to be used with their proton synchrotron, the optical pumping approach is planned for the polarized H^- source. (Y. Mori, this volume)

4. POLARIZED ELECTRON SOURCES

Although sources of polarized electrons[14,15] was not a topic at our workshop (it probably should be at any future such workshop), it may be useful to record here that two types of polarized electron sources have been successfully used in high energy experiments, both with the 20 GeV accelerator at SLAC. The first (Fig. 4) is based on photoionization of an electron spin-polarized ^{6}Li atomic beam[16], and the second (Fig. 5) on photoemission of spin-polarized electrons from GaAs with circularly polarized laser light.[17] Operating conditions of these sources are given in Table VI. Some of the problems for polarized electron sources, such as production of an intense polarized atomic beam and laser technology, are similar to those for polarized proton sources.

5. NEW ACCELERATOR-RELATED PROJECTS

Several relatively new projects for sources of polarized protons for accelerators are listed in Table VII. Although this list is probably incomplete, it indicates that substantial new sources have been recently completed or are being developed.

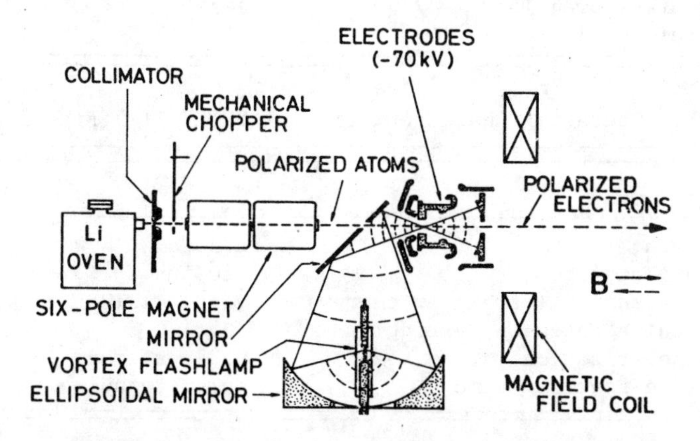

Fig. 4. Schematic drawing of the polarized electron source PEGGY showing the principal components of the lithium atomic beam, the uv optics, and the ionization region electron optics.

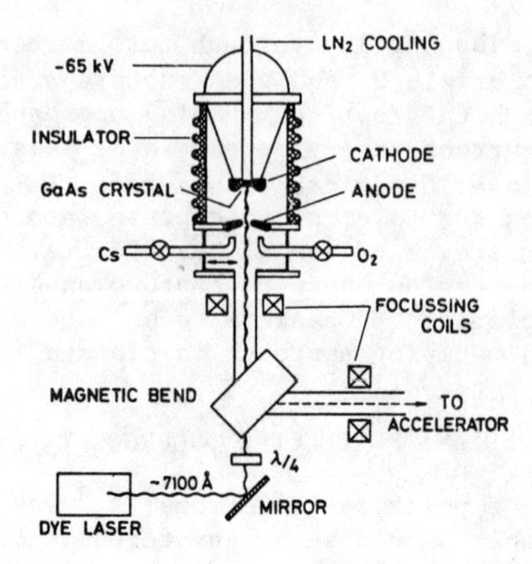

Fig. 5. Schematic diagram of the GaAs polarized electron source at SLAC.

Table VIa. Operating Characteristics of PEGGY I (Fig. 4)

Characteristic	Value
Pulse length	1.6 μs
Repetition rate	180 pps
Electron intensity (at high energy)	~10^9 e^-/pulse
Pulse to pulse intensity variation	<5%
Electron polarization	0.85±0.07
Polarization reversal time	<1 s
Time between reversals	2 min
Intensity difference upon reversal	<5%
Lifetime of Li oven load	175 h
Time to reload Li	43 h

Table VIb. Operating Characteristics of PEGGY II (Fig. 5)

Characteristic	Value
Pulse length	1.5 μs
Repetition rate	120 pps
Electron intensity (at higher energy)	(1 to 4)x10^{11} e^-/pulse
Pulse to pulse intensity variation	~3%
Electron polarization	0.37, average
Polarization reversal time	pulse to pulse

Table V11. Relatively New Projects

High Energy

(1) Saturne, Saclay; 3 GeV
Ground state H^+ source; operating

(2) AGS, Brookhaven; 30 GeV
Ground state H^- source using Cs beam ionization; under development

(3) KEK, Tsukuba; 12 GeV
Optically pumped polarized Na; under development

(4) SPS, CERN; 400 GeV
Polarized H,D jet target using ground state H
Experiment approved using recirculating H^+ beam

(5) FNAL; 500-1000 GeV
Polarized p beam from Λ decay
Beam line and experiments approved

Low Energy

Many improvement programs
Yale, Tandem electrostatic accelerator, 22 MV.
Ground state H^- source planned

6. CONCLUSIONS

The need in physics for high energy beams of polarized protons and deuterons appears compelling. Because of the high cost of accelerator operation, it is of the greatest importance to improve the quality of our polarized sources. Quantitative understanding can no doubt be helpful, so that, for example, better knowledge of various atomic cross sections, depolarizing mechanisms, flow conditions in atomic beam formation, and velocity distributions in atomic beams are needed. An entire polarized source with its many closely coupled sections should be designed as a unit. Ideally, as for unpolarized high energy beams, the intensity limit for polarized beams should be determined by the acceptance and space-charge effects in the accelerator.

This work was supported in part by the U.S. Department of Energy under Contract No. DE-AC02-76ER03075.

REFERENCES

1. High-Energy Physics with Polarized Beams and Polarized Targets, edited by C. Joseph and J. Soffer (Birkhäuser Verlag, Basel, 1981).
2. Polarization Phenomena in Nuclear Physics - 1980, edited by G.G. Ohlsen, R.E. Brown, N. Jarmie, W.W. McNaughton, and G.M. Hale (American Institute of Physics, New York, 1981).
3. Isabelle, Proceedings of the 1981 Summer Workshop, BNL Report No. 51443, 1981, Vol. 2, p. 601.
4. Proceedings of the International Symposium on Polarization Phenomena of Nucleons, edited by P. Huber and K.P. Meyer (Birkhäuser Verlag, Basel, 1961).
5. S. Dushman, Scientific Foundation of Vacuum Technique, 2nd ed., edited by J.M. Lafferty (Wiley, New York, 1962).
6. N.F. Ramsey, Molecular Beams (Clarendon Press, Oxford, 1956).
7. P. Kusch and V.W. Hughes, "Atomic and Molecular Beam Spectroscopy," Encyclopedia of Physics XXXVII/1, edited by S. Flugge (Springer-Verlag, Berlin, 1959).
8. H. Lew, in Methods of Experimental Physics, edited by V.W. Hughes and H.L. Schultz (Academic Press, New York, 1967), Vol. 4, Part A, p. 155.
9. H.S.W. Massey, Electronic and Ionic Impact Phenomena (Clarendon Press, Oxford, 1971), Vol. 3.
10. M.E. Gersh and R.B. Berstein, Chem. Phys. Lett. 4, 221 (1969).
11. H.F. Glavish, in Higher Energy Polarized Proton Beams, edited by A.D. Krisch and A.J. Salthouse, AIP Conference Proceedings #42 (American Institute of Physics, New York, 1978), p. 47.
12. S.B. Crampton, T.J. Greytak, D. Kleppner, W.D. Phillips, D.A. Smith, and A. Weinrib, Phys. Rev. Lett. 42, 1039 (1979).
13. L. Dick, J.B. Jeanneret, W. Kubischta, and J. Antille, High-Energy Physics with Polarized Beams and Polarized Targets, edited by C. Joseph and J. Soffer (Birkhäuser Verlag, Basel, 1981), p. 212; T.O. Niinikoski, ibid, p. 191; L. Dick (private communication, 1980).
14. J. Kessler, Polarized Electrons (Springer-Verlag, Berlin, 1976).
15. G. Baum, in Polarization Phenomena in Nuclear Physics - 1980, edited by G.G. Ohlsen, R.E. Brown, N. Jarmie, W.W. McNaughton, and G.M. Hale (American Institute of Physics, New York, 1981), p. 785.
16. M.J. Alguard, J.E. Clendenin, R.D. Ehrlich, V.W. Hughes, J.S. Ladish, M.S. Lubell, K.P. Schüler, G. Baum, W. Raith, R.H. Miller, and W. Lysenko, Nucl. Instrum. Methods 163, 29 (1979).
17. C.Y. Prescott, W.B. Atwood, R.L.A. Cottrell, H. De Staebler, E.L. Garwin, A. Gonidec, R.H. Miller, L.S. Rochester, T. Sato, D.J. Sherden, C.K. Sinclair, S. Stein, R.E. Taylor, J.E. Clendenin, V.W. Hughes, N. Sasao, K.P. Schüler, M.G. Borghini, K. Lübelsmeyer, and W. Jentschke, Phys. Lett. 77B, 347 (1978).

LAMB-SHIFT POLARIZED ION SOURCES---AFTER 15 YEARS

Thomas B. Clegg
Department of Physics, University of North Carolina*
Chapel Hill, North Carolina 27514
and
Triangle Universities Nuclear Laboratory*
Duke University
Durham, North Carolina 27706

ABSTRACT

The technique of producing spin-polarized beams of H^- and D^- ions by the Lamb-shift method is reviewed. Maximum useable beam intensities are shown to depend on the flux of metastable $H_0(2S)$ or $D_0(2S)$ atoms which can be obtained with a useable beam emittance. Factors limiting this flux are discussed. Two methods of spin-state selection, the spin-filter, and the diabatic field-reversal process, are discussed. The former selects atoms in a single nuclear hyperfine state and is the best method for $\vec{D}^-$ ions; the latter yields larger beam intensities and is often the best choice for $\vec{H}^-$ ion beams. Best sources of the Lamb-shift type produce currents ≥1μA with proton beam polarizations ≥0.7. Intensity increases of up to a factor of 3 may be possible by improving the present technology, but progress will be difficult to achieve.†

INTRODUCTION

The Lamb-shift polarized ion source is based on ideas first proposed by Lamb and Retherford[1] to polarize the $H_0(2S)$ atom by selective quenching of atoms in some of the hyperfine states to the $H_0(1S)$ ground state. Madansky and Owen[2] first proposed using a charge-exchange reaction to produce an $H_0(2S)$ atomic beam, but it remained for Donnally and Sawyer[3] to propose the first practical technique for obtaining polarized ion currents of useable intensity. Since the first such ion source was installed on an accelerator in 1967,[4] subsequent work in a large number of laboratories has developed the technology extensively. Lamb-shift sources are installed on approximately 15 accelerators worldwide. In addition parts of the source producing the metastable $H_0(2S)$ atomic beam are used on at least two experiments to search for parity non-conservation in atomic hydrogen.[5,6]

Reviews of the Lamb-shift technique have appeared.[7-12] Recently these have made it evident that progress in developing Lamb-shift source technologies has slowed, and evaluation of the existing, mature, Lamb-shift source design must be made in detail to determine the potential for further improvements. That will hopefully be attempted here at this workshop.

The schematic of the Lamb-shift technique is shown in Fig. 1. Metastable beams of $H_0(2S)$ (or $D_0(2S)$) are obtained by charge exchange of a 550 eV (1100 eV) H^+ (D^+) beam in cesium vapor. Metastable atoms emerging from the cesium charge-exchange canal are separated from unwanted charged ions with small transverse electric fields (≤15 V/cm) before they enter an axial magnetic field of ~575 G. Here, with electric fields, atoms in one or more nuclear hyperfine states may be selected by causing the other unwanted atoms to decay to the ground state. The resulting nuclear-polarized $H_0(2S)$ (or $D_0(2S)$) beam is then ionized selectively by charge exchange in argon gas. Argon is chosen because of its much lower cross section for producing H^- (or D^-) ions from unpolarized ground-state atoms.[3] One may also obtain $\vec{H}^+$ (or $\vec{D}^+$) ions from a Lamb-shift source by selective charge exchange in iodine,[13] but the output currents are not competitive with the ~100 μA now available from atomic-beam sources.[14]

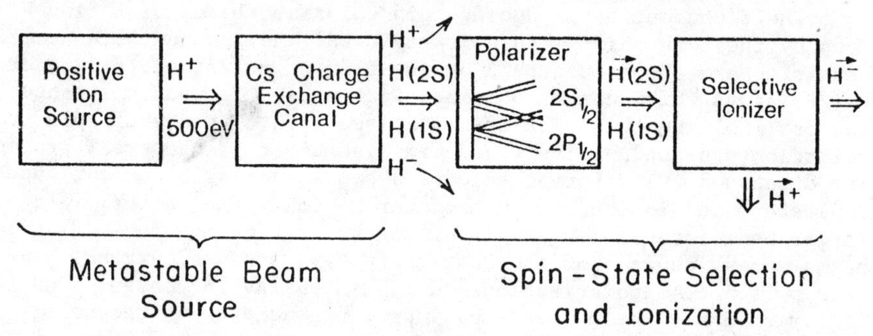

Fig. 1. Schematic showing the basic components of a Lamb-shift polarized ion source.

The Lamb-shift source has basically two important and rather independent sections. First, the positive ion source and cesium charge-exchange region comprise the source of metastable atomic beam and are critical to the output beam intensity. Then, the polarization scheme chosen and the ability to reverse the spin quantization axis rapidly before ionization affect most the versatility of the source for experiments. In the material to follow, these two sections of the source will be studied in detail.

SOURCES OF METASTABLE ATOMS

Sources of Positive Ions

There are two different philosophies for forming the H^+ (or D^+) positive ion beam needed.[9] There are "extended geometry" sources in which the H^+ (or D^+) beam is extracted at an energy higher than required and allowed to expand radially before refocussing and decelerating through the cesium canal. The goal here is to minimize the space-charge forces by expanding the beam quickly at a higher energy before decelerating through the cesium canal with a large (≥1 cm diameter) converging beam. Such a system is used at Munich[15] with an r.f. source for the positive ions. At Giessen[16] a duoplasmatron is used. The r.f. source operates quite reliably, but the duoplasmatron may be preferable because of its higher gas efficiency and lower output-beam energy spread of 20 eV for H^+ and 12 eV for D^+ ions.[17] The "extended geometry" source is very sensitive to alignment, and because it utilizes electrostatic focussing, prohibits capturing negative ions or electrons in the beam for space-charge neutralization before deceleration.

The second style of positive ion source uses the "close-coupled" geometry first introduced by J. L. McKibben at Los Alamos.[9] Here the beam is extracted at energies up to 6 to 8 keV through a 6 to 9 mm diameter hole and decelerated immediately to enter the cesium canal. The beam stays small in diameter but only for a short time to minimize beam blowup from internal space-charge forces. It is felt that cesium from the nearby canal also is ionized by the beam, producing many slow electrons to facilitate beam space-charge neutralization.

The duoplasmatron is the most common source of positive ions. If one looks in detail at those used in several labs, one finds both similarities and differences. The geometries of the anode aperture and plasma expansion cup region for several Lamb-shift sources are compared in Fig. 2. Both typical designs shown include a cooled copper

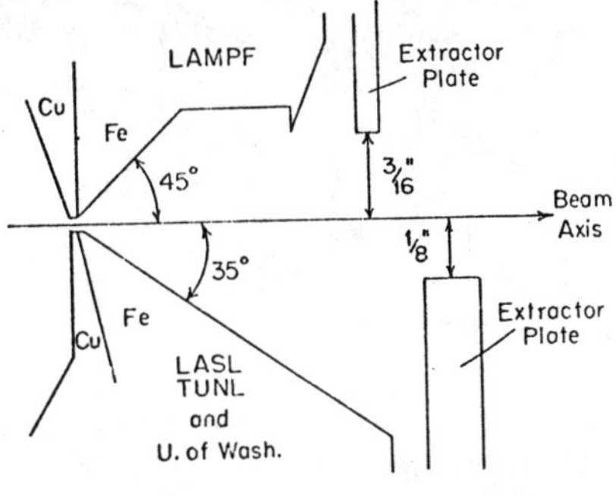

Fig. 2. Schematic showing the duoplasmatron anode-aperture and extraction regions of Lamb-shift sources. Above the horizontal axis is the design used at LAMPF. This is similar to the cylindrical-cup geometry used also at Giessen. Below the horizontal axis is the conical expansion cup design used in several laboratories. In both designs a cooled cooper aperture removes heat generated by the arc and the magnetic aperture shields the expansion cup region from large magnetic fields.

aperture to remove the power generated by the arc and a steel aperture to attenuate the magnetic field from the arc region so it is as low as possible in the plasma expansion cup. At LAMPF the cylindrical expansion cup with a knife-edge rim is used to stabilize the surface from which beam is extracted during pulsed operation. At Giessen it has been found that the output beam intensity peaked sharply when the ratio D/L of the anode-aperture-channel diameter to its length was varied.[18] Similar critical dependence of the beam current on these dimensions has been found at our laboratory, but the cause is not completely understood.[19]

Cesium Oven Designs

Three different styles of cesium oven are used in Lamb-shift sources. At TUNL,[20] at Giessen, and several other laboratories the oven is a heated reservoir connected through a heated valve to a heated charge-exchange canal. The canal is usually ~15 cm long and ~1.3 cm diameter and the cesium density in the canal is $\sim 3.5\times10^{13}$ atoms/cm^3 at the operating temperature of ~100°C. Cesium consumption rates from this type of canal, from loss out the ends, are ~1 gm/day.

A second cesium oven design is used at Munich[21] and is shown in Fig. 3. During operation the reservoir is kept at T = 200°C and a

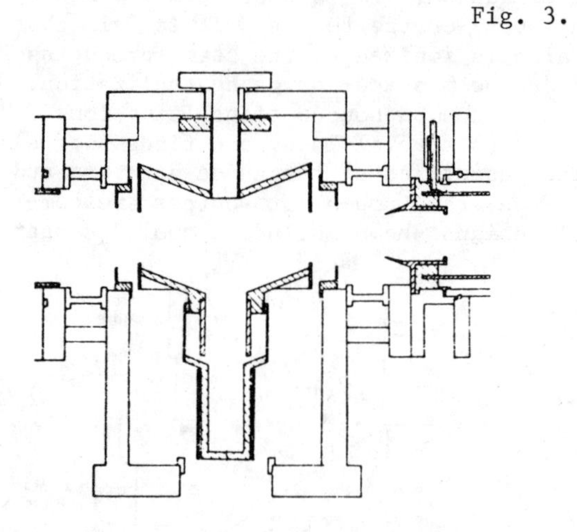

Fig. 3. The recirculating cesium oven used at the Munich laboratory is shown. The oven is 12 cm long with apertures on each end of 3.4 cm diameter. The whole oven is cooled to 0°C when not in use.

jet of cesium is evaporated upward across the beam before colliding with a surface at 35°C where it condenses and falls back into the reservoir. This recirculating oven design is more efficient than the first type described; a cesium loss rate of 2.4 gm/day is observed even though the charge-exchange canal is shorter and the area for loss of cesium out the ends is ~7 times larger.

The most efficient oven design was developed at Los Alamos by Hardekopf.[22] The oven is basically like the first design described except a stainless steel mesh is present inside the canal to act as a wick for molten cesium. During operation the center of the canal is heated to ~80°C while the ends are cooled to 10 to 20°C. Cesium atoms, which vaporize near the center of the canal and then collide with the cold end walls of the canal, stick and are drawn back toward the canal center by capillary action in the wick. Details of the construction and use of such a wicked oven are given by Hardekopf.[22] At the University of Washington, Risler and Trainor report measurements of the cesium flux out the ends of such a wicked canal using hot-wire ionization to detect the cesium.[23] They measure a loss of ~0.1 gm/day from a 15 cm long, ~1.3 cm diameter canal with the center of the canal at 100°C. Thus this oven is the most efficient in confining the cesium and reducing cleanup problems. It has the additional advantage that the presence of cesium in the wick allows stable optimum production of 2S metastable atomic beams within only 5 minutes after beginning to heat the oven and canal. As with either of the other two oven designs discussed above, the most reliable, stable operation is obtained when the cesium oven and canal can be isolated with valves and kept under vacuum while maintenance is underway on other parts of the ion source.

Metastable Beam Controls and Diagnostic Measurements

It has often been found that optimization of the neutral, metastable atomic beam required an ability to move the electrodes within the positive ion source, primarily to steer the charged ions as they enter the cesium canal. At the University of Washington, Holmgren and Trainor have installed a steering magnet in front of the cesium canal.[6] Then 3.3 m away at the end of their apparatus,[6] they have installed a secondary-electron detector for the metastable atomic beam, which has four sections to detect the up-down and left-right motions of the atomic beam. Their magnet and detector designs are shown in Fig. 4. The up-down and left-right detector signals are

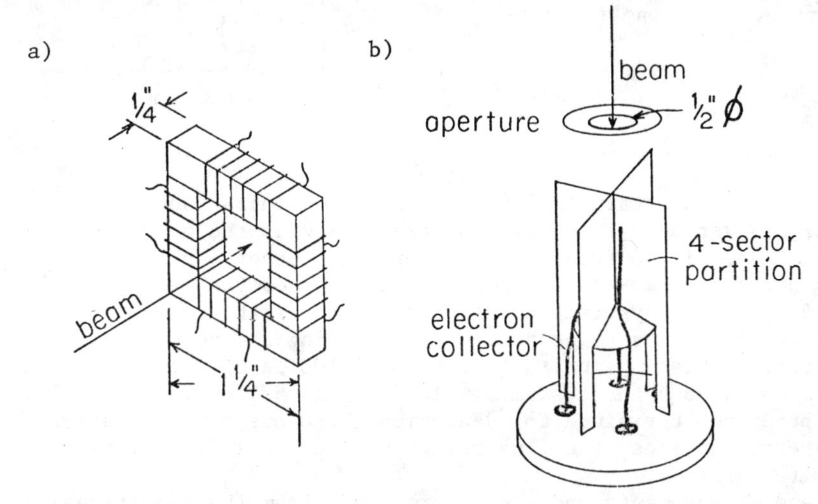

Fig. 4. a) Shown is the "picture frame" magnet used at the University of Washington, with each of the four sides of the magnet made of copper wire wound on ferrite. The magnet is capable of 25G maximum field at its center. b) The 4-sector secondary-electron detector collimates the entering metastable beam to impinge on a conical surface where the secondaries are produced.

processed to obtain a difference signal which is then summed with the constant d.c. currents passing through the steering magnet coils. This feedback stabilizes the metastable beam position passing through the system, allowing an overall noise reduction of -40 db for frequencies of 0-100 Hz and -6 db for 10 kHz.[24]

Metastable beam intensity measurements have been attempted at the Triangle Universities Nuclear Laboratory using an ionization chamber to detect the Lyman $-\alpha$ radiation emitted when the 2S metastable atoms are quenched to the ground state.[25] Fluxes were measured for $H_0(2S)$ and $D_0(2S)$ atoms in a single hyperfine state at a point 88 cm beyond the end of the cesium canal, and the results are shown in Fig. 5. Intensities quoted are accurate only to $\pm$ 50%

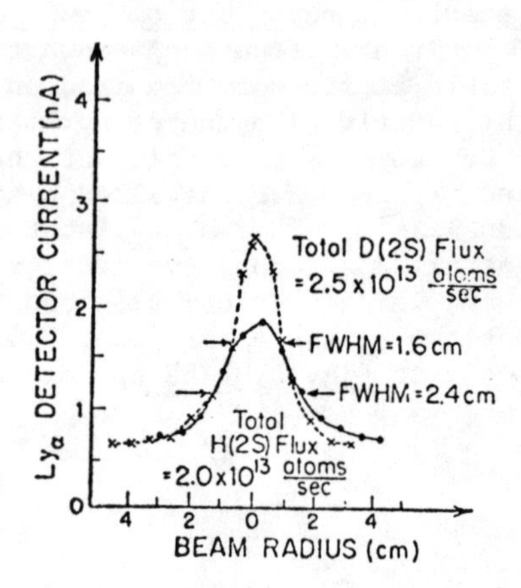

Fig. 5. Metastable atomic beam flux measurements made in the Lamb-shift polarized source at the Triangle Universities Nuclear Laboratory. The results shown were made with the spin filter operating and are for a single hyperfine state. Later measurements gave fluxes both for $H_o(2S)$ and $D_o(2S)$ which were four times larger than shown here with the same overall beam profiles.

because of the uncertainty of calibrating the ionization chamber detector. Nevertheless, these results and indeed the output polarized beam intensities from other Lamb-shift sources, indicate that between 8×10^{13} and 1.9×10^{14} atoms/sec emerge from the cesium canal in all four $H_0(2S)$ hyperfine states having an emittance suitable for conversion to $\vec{H}^-$ ions to be accelerated. Using the results of Schlachter,[26] this implies that from 40 to 100 μA of useable H^+ ion beam passes through the cesium canal. Roughly 50% more $D_0(2S)$ flux can be produced, but since for deuterium there are six instead of four hyperfine states, the flux per hyperfine state is nearly the same as for protons.

These measurements and simple arguments from the geometrical dimensions of apertures inside several Lamb-shift sources[15,16] imply a metastable atomic beam emittance of less than 3π mm-mrad-MeV$^{1/2}$. Measurements of accelerated $\vec{H}^-$ beam emittance made at LAMPF[27] and TRIUMF[28] indicate that subsequent ionization and acceleration to several hundred keV largely preserves this excellent beam quality.

The metastable-beam flux-measuring system at TUNL allowed also the determination that the conversion efficiency in argon gas for producing $\vec{H}^-$ ions from $H_0(2S)$ atoms is ~11%.

Metastable Beam Sources --- Conclusions

Extensive effort at several laboratories has failed recently to improve metastable beam intensities. At present there is no clear choice between "extended geometry" and "close-coupled" extraction systems for positive ions, since the systems are capable of comparable performance.[27,29] Space-charge neutralization of the beam is extremely important, both to avoid radial divergence of the slow, positive ion beam because of internal space-charge forces and to avoid quenching of the $H_0(2S)$ atoms emerging from the cesium charge-exchange canal. This "self-quenching" of the metastable atoms by electric fields within the beam has been observed in sources at TUNL and Los Alamos which use the "close-coupled" extraction geometry to form the positive ion beam. At present it is not clear whether this observation indicates only that one must be more careful to neutralize space charge or whether this problem will place an ultimate limit on the metastable atomic beam intensity. In either case, by extending present technologies and paying careful attention to detail, further increases of a factor of three in intensity may be possible.[†]

SPIN-STATE SELECTION AND IONIZATION

Two methods of obtaining nuclear polarization, the nuclear spin filter and the diabatic field reversal scheme are used with Lamb-shift sources. Both are well understood and operate quite reliably.

The Nuclear Spin Filter

First developed at Los Alamos[30] the spin filter requires an r.f. cavity operating inside a very uniform 575 $\pm$ 0.2G axial magnetic field. If d.c. and r.f. electric fields are applied simultaneously to the 2S metastable beam while it is inside this magnetic field, a three-level resonance process can be established[31,32] which allows one to preserve 2S metastable atoms in a single hyperfine state while quenching atoms in all other hyperfine states to the ground state. The resonant levels for $D_0(2S)$ are shown in Fig. 6a. The selection of a single hyperfine state is nearly 100% efficient,[25] and switching from one hyperfine state to another to change the beam polarization is accomplished by switching between hyperfine level crossings with changes in the spin-filter magnetic field. The method is most effective for deuterium beams, since field changes of only $\pm$ 10G are required to switch between purely tensor polarized and purely vector polarized beams.[33] This can be accomplished by rapid switching with an auxiliary, axial-field coil surrounding the spin-filter cavity as first suggested by Ohlsen[34] and now implemented at Tsukuba.[35] Typical resulting beam currents and polarizations are shown in Fig. 6b.

A real advantage of the Lamb-shift source, when used on accelerators where the beam polarization is not diminished during the acceleration process, is the possibility of a "quench-ratio" measurement to determine the beam polarization.[36] This requires a simple measurement of the ratio of beam intensity when the spin-filter is tuned to select one of the hyperfine states to the intensity (the plotted line in Fig. 6b) when a strong electric field is applied to the metastable atomic beam to quench all 2S atoms fo the 1S ground state. If this ratio is Q, and if one assumes that all the beam selected in a single hyperfine state is completely polarized while all background beam (below the dotted line) is completely unpolarized, then the beam polarization is given by $P_b = 1 - 1/Q$. Rapid beam polarization measurements via the "quench ratio" with an accuracy of 3% to 5% are routine and extremely easy. An accuracy of ~1% is possible,[36] but requires considerable care by the experimenters. The quench ratio does not reveal, for example, whether the 2S atoms were ionized in the argon region in a large enough magnetic field to align the electron magnetic moment in the "$M_I = 0$" and "$M_I = -1$" states, nor does it reveal whether the spin quantization axis has been precessed to the proper direction at the target. Even so, it often provides in high-energy proton experiments one of the most convenient and accurate ways to determine the beam polarization.

a) 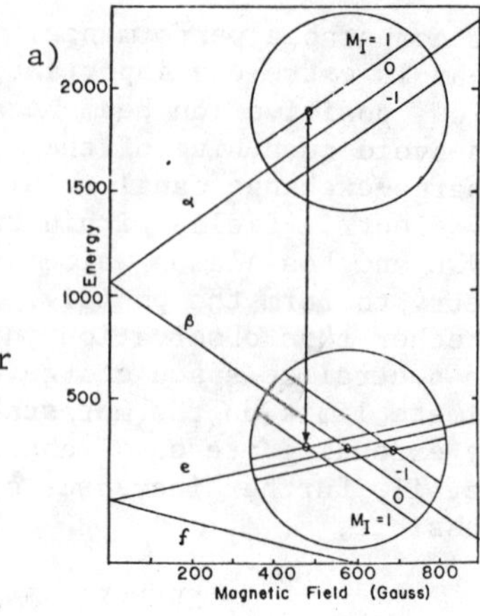

b) 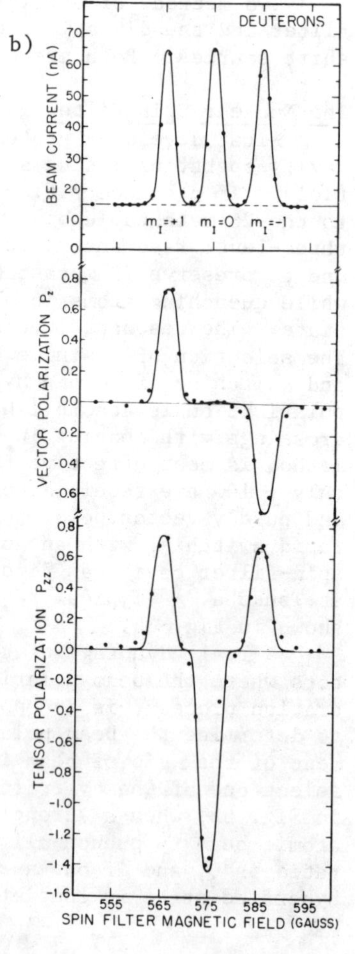

<u>The Diabatic Field Reversal Method</u>

The second nuclear polarization scheme was proposed first by P. G. Sona[37] and often bears his name. It requires that the 2S metastable atomic beam enter a region of axial 575G magnetic field where, as shown schematically, at axial position a in Fig. 7, the atoms in the lower hyperfine states are removed from the beam by a small ~15 V/cm electric field which quenches them to the ground state. The remaining 2S atoms emerge from this 575G field into a region where the axial field is reversed. Here, if

Fig. 6. a) A Zeeman energy level diagram for the 2S and 2P states of atomic deuterium showing in the enlarged regions the hyperfine level crossings. The vertical arrow indicates the three levels involved in the resonant selection of the $M_I = +1$, α-state by the spin filter. b) Measured values of beam current, P_z, and P_{zz} for the polarized deuteron beam emerging from a spin filter. A change of ±10G in the axial magnetic field is all that is necessary to switch from one hyperfine state to another. The polarization values shown resulted from argon ionization in ∿40G magnetic field.

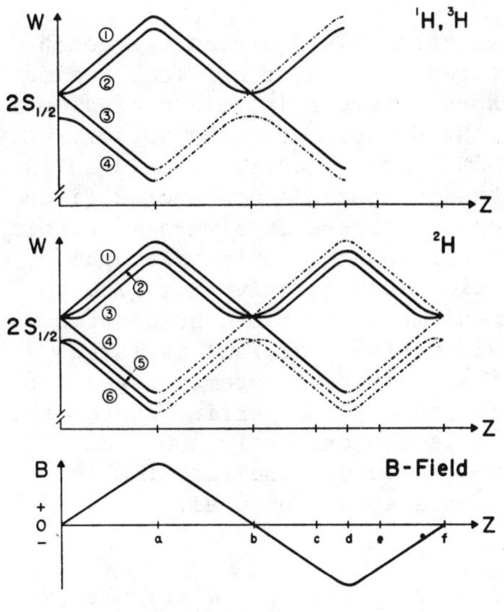

Fig. 7. Shown plotted versus Z are the Zeeman splitting for ^{1}H, ^{2}H, and ^{3}H and the qualitative axial magnetic field. Dotted lines show states quenched at a and d. If a diabatic field reversal occurs at b, then ionization in argon at different axial positions Z gives the theoretical polarization values shown.

IONIZATION AT Z=	PROTON & TRITON POLARIZATION |P|	DEUTERON POLARIZATION $\|P_3\|$	DEUTERON ALIGNMENT P_{33}
b	1/2	1/3	-1/3
c	1	2/3	0
e	—	1/2	-1/2
f	—	0	-1

the local magnetic field which any individual atom experiences is caused to flip in axial direction fast by comparison with the atom's Larmor precession period in that field, then atomic spin will not follow the field reversal. Thus atoms in state 1 (see Fig. 7) will have their spin projection reversed with respect to the external field. If they then enter a stronger region of this reversed B-field (at position c in Fig. 7) where they are ionized in argon, then one obtains $\vec{H}$ beams with maximum nuclear polarization or $\vec{D}$ beams with pure vector polarization. Deuteron beams with pure tensor polarization result if the lower hyperfine state is quenched in a 575G field at axial position d. The remaining 2S atoms are ionized in a weak field at position e.

A real advantage of this "Sona scheme" is that atoms in two or three hyperfine states are used rather than one, thereby increasing the available beam current by a factor of 2 or 3 over that with the spin filter. For this increase one sacrifices the ability to make a quench-ratio beam-polarization determination, since only for hyperfine state 1 is the polarization independent of the magnetic field in which the beam is ionized. In high energy proton acceleration where there may be some resonant depolarization during accelerators which makes the quench-ratio measurement meaningless, the diabatic field reversal scheme is clearly preferable.

Hardware to implement the diabatic field reversal is a very simple arrangement[16] of two axial magnetic field coils to produce the $\pm$ 575G regions and the enclosed electric field plates to quench 2S atoms in the unwanted hyperfine states. The most critical region occurs where the magnetic field reverses. Here axial field gradients of $\leq$0.5G/cm are needed to accomplish the diabatic reversal effectively for atoms at all beam radii $\leq$1 cm. Sometimes transverse field coils are added in this region to compensate any spurious transverse field which would destroy the diabatic reversal. There is always an effort to make these systems occupy the shortest possible axial distance by using iron plates to reduce the magnetic field quickly from 575G to near zero. The danger is that the axial magnetic field gradient becomes so large that the radial field $B_r = -(r/2)(dB_z/dz)$ is large enough to cause quenching of the desired metastable atoms in a $\vec{v} \times \vec{B}$ motional electric field. For atoms in the upper hyperfine states the allowable maximum axial field gradient is approximately 100 G/cm. For atoms in the lower hyperfine states, however, the lifetime is considerably shorter and smaller gradients should be used.

Rapid Spin Reversal Processes

The use of beams from Lamb-shift sources for very difficult experiments searching for parity non-conservation in nuclear physics[38,39] stimulated the development of schemes for rapid reversal of the spin quantization axis. The schemes used most often were developed first by J. L. McKibben,[40] and require that the metastable beam be prepared in a single hyperfine state using a spin filter before reversal occurs. Two schemes used most often will be described here.

The first rapid-reversal process is shown schematically in Fig. 8a. Here the magnetic fields in the spin filter and argon regions of the source are opposite to one another so normally the entering beam in hyperfine state 1 undergoes a Sona transition at the field reversal point. If, however, at this point a carefully designed transverse B-field is applied such that in the region of the field reversal the magnitude of the B-field is held roughly constant at ~2G while the direction of the B-field reverses, one can arrange that the metastable atom's spin precesses once during the field reversal and follows B to emerge with the spin quantization axis reversed.[41] Then a rapid spin flip is accomplished simply by switching on and off the transverse B-field.

The hardware used to accomplish this is shown in Fig. 8b. Four axial-field coils and one transverse-field coil shape the B-field to have the configuration shown in the graph. To compensate for any quenching which might occur in the motional $\vec{v} \times \vec{B}$ electric field when the transverse field is applied, additional plates are used to make an appropriate compensating transverse E-field. With this system operating correctly, spin reversal can be accomplished at ~1kHz with a beam current modulation of 1 part in 10^4 and beam motion on target of less than 3×10^{-6}cm and 1×10^{-7} radians.[41]

The second rapid-reversal process has been used by McKibben[40] and is soon to be installed on Lamb-shift sources at LAMPF[27] and TUNL. The hardware for this scheme is shown in Fig. 9. The reversal is accomplished by reducing the axial magnetic field completely to

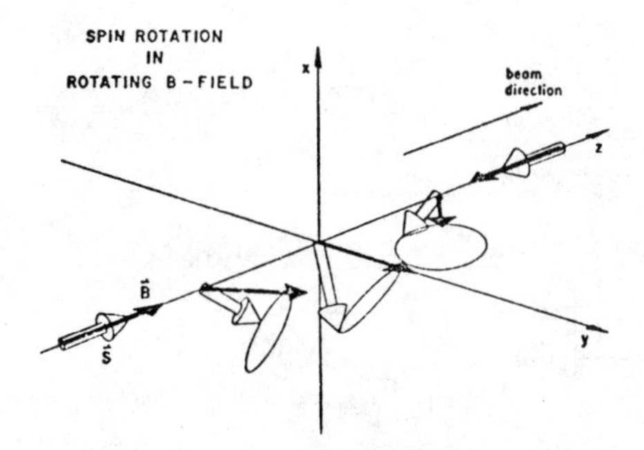

Fig. 8 a) Rapid spin reversal can be accomplished by turning on or off a small transverse magnetic field which either destroys or enables a diabatic field reversal.

Fig. 8 b) The system of magnetic field coils and the resulting magnetic field to accomplish the rapid spin reversal. The transverse field coil (T-coil) is wound on a frame which is either non-conducting or split into sections to minimize eddy currents and maximize the rate at which the transverse field can be switched.

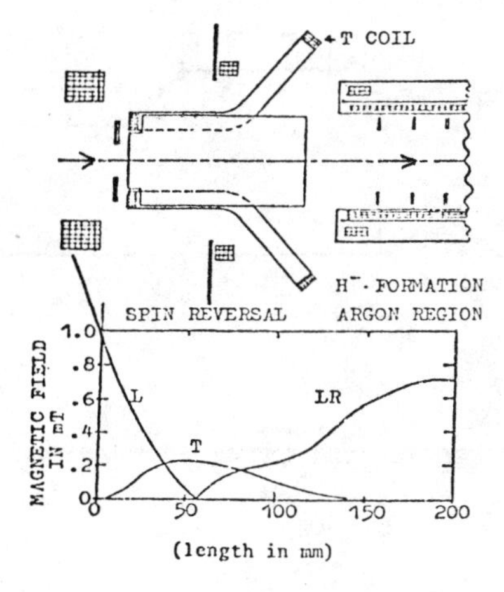

approximately 0.01G following selection of hyperfine state 1 in the spin filter. Then in this zero-field region a small transverse B-field is applied around which the spin precesses. If the integral of this transverse field experienced by the beam is ~11 G-cm, precession through +90° results. Reversing the direction of this transverse field reverses the precession to -90°. Rotation of this transverse field around the beam axis rotates the plane in which the spin precesses. Thus, this scheme accomplishes not only rapid spin reversal by rapid reversal of this small transverse B-field, but the rotatable feature acts also as a spin precession device to facilitate any arbitrary orientation of spin quantization axis. A small transverse electric field can be added orthogonal to the B-field to cancel any $\vec{v} \times \vec{B}$ motional electric field which would quench the beam. After this spin precession the beam must emerge from the transverse B-field region to be ionized in argon in zero magnetic field, otherwise further spin precession would occur.

The hardware for this device has two critical components. First one must construct a system of magnetic shields or coils to reduce the residual magnetic field to ~0.01G throughout the fast-flip and argon regions. The system being constructed at TUNL to do this is shown in Fig. 9a. Second, the transverse field coil must produce as small a region of axial fringing field as possible to avoid precession to unwanted orientations. To assure that all atoms in the beam

a)

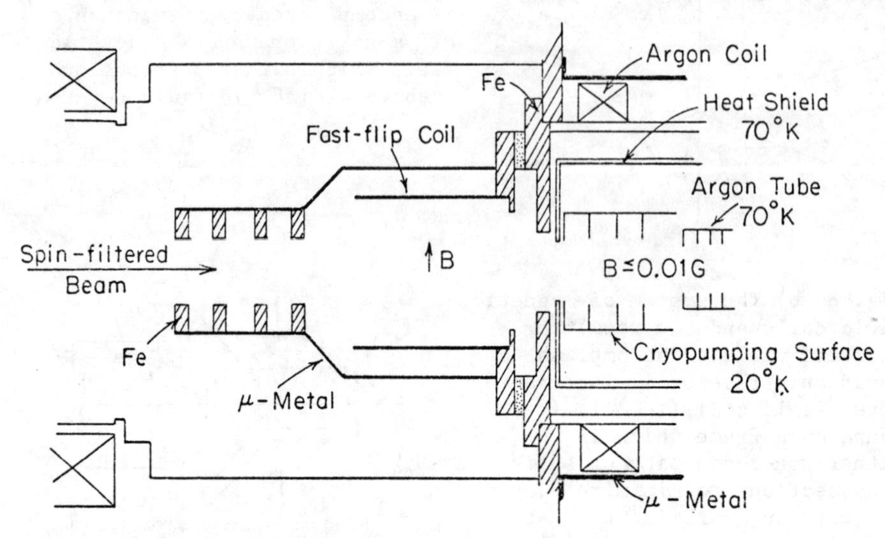

b)

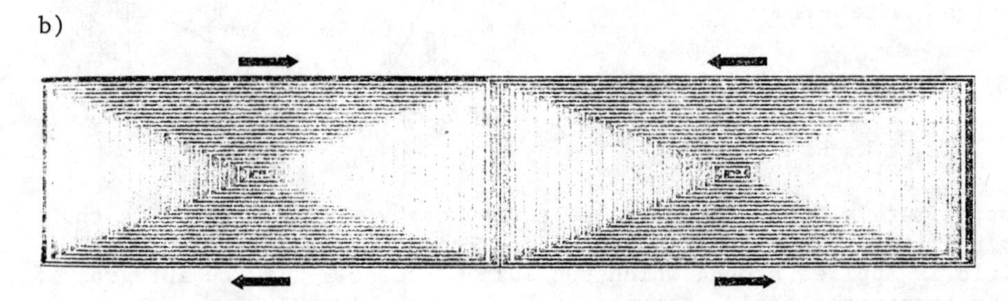

c)

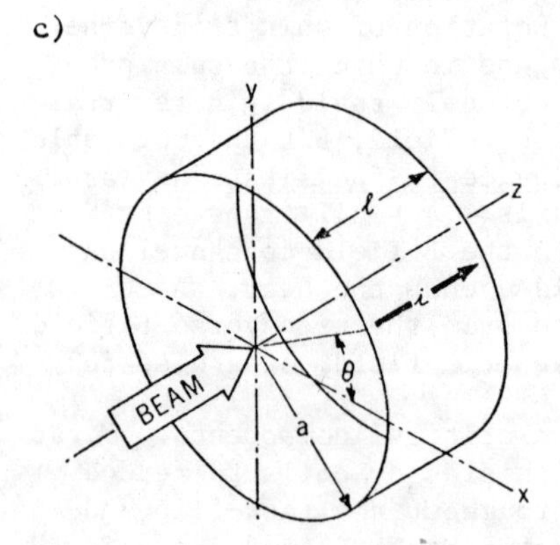

Fig. 9 a) Hardware under construction at the Triangle Universities Nuclear Lab to accomplish rapid spin-axis reversal by reducing the magnetic field to zero and then applying a small transverse B field around which the spin will precess. b) Printed circuit board layout which produces a transverse dipole field for which the integral of $\vec{B}\cdot d\vec{\ell}$ is constant for a large portion of the beam. c) Desired configuration of the circuit-board dipole-field coil with the beam passing along the axis of the cylinder.

precess by the same amount, it is important that the integral of $\vec{B} \cdot d\vec{\ell}$ experienced by each part of the beam is the same. A system which seems to accomplish this conveniently is based on printed-circuit coils[42] of the type shown in Fig. 9b. When wound into a cylinder around the beam axis as in Fig. 9c, they produce a dipole magnetic field with the properties desired. At TUNL D. A. Kopf is fabricating two sets of these coils, one on the inside and one on the outside of a cylinder, by using double-sided circuit board. If the dipole fields from the inside and outside sets are orthogonal, then the currents through these coils can be made proportional to sin ϕ and cos ϕ, respectively, where ϕ is an angle measuring orientation around the beam axis. Superposition of these two orthogonal, transverse fields produces a field B whose magnitude is constant and whose orientation can be controlled simply by turning a sine-cosine potentiometer.[42a]

To be as versatile as possible this last fast-flip hardware should be compatible with operation, without the capability of a fast spin-axis reversal, 1) to utilize the Sona-scheme for obtaining nuclear polarization and the highest possible intensity for operation with $\vec{H}^-$ beams; or 2) to accommodate spin-filter operation of the sources for $\vec{D}^-$ beams in hyperfine states 2 and 3. Both of these features require that one maintain the capability of an axial B-field of up to 200 G in the argon region to obtain the full nuclear polarization of the beam.

Selective Argon Ionization and Acceleration

Although extensive searches have been made,[43] no charge-exchange medium has been found preferable to argon for selectively producing $\vec{H}^-$ ions from $H_0(2S)$ atoms. The process must, of course, discriminate against producing H^- ions from $H_0(1S)$ atoms in the beam. In modern Lamb-shift sources, and especially most of those utilizing fast spin-axis reversal, the argon is highly localized spatially by cryopumping to avoid charge-exchange occurring in unwanted regions. A closed cycle helium refrigerator which produces ~2W of cooling at 20°K is suitable for this purpose. Usually, 20°K baffles are located close to both ends of the argon canal, and a 70°K surface surrounds this to act as a heat shield, to cool the incoming gas and cool the argon canal.

Following ionization most sources accelerate the $\vec{H}^-$ (or $\vec{D}^-$) beam rapidly to form a small beam waist. It has been found[44] that a small ~2 mm diameter aperture at this waist serves to enhance the beam polarization. This occurs because H^- ions produced via argon charge-exchange from $H_0(1S)$ atoms do so with larger momentum transfer than do $\vec{H}^-$ ions originating from $H_0(2S)$ atoms. This leads to larger transverse momenta and thus larger beam emittance for these unpolarized, background H^- ions. Removing them by "scraping" off this unpolarized beam halo is best accomplished as soon as possible after ionization.

Spin-State Selection and Ionization - - - Conclusions

The Lamb-shift source should be adapted to the experimental program where it will be used. Clearly if only $\vec{H}^-$ ions are needed and if there is likely to be depolarization during acceleration, then a simple source employing the diabatic-field-reversal scheme is best. For the same beam polarization it yields ~2.2 more beam intensity[41] than does a source with a spin filter. If quench-ratio beam polarization measurement or $\vec{D}^-$ operation are important, then the spin filter must be the choice. If, however, as on many lower energy tandem accelerators one desires the good features of both of these modes of operation as well as the facility for rapid spin-axis reversal, then one should choose a hybrid source. It should include a spin filter followed by a region which can either facilitate rapid spin-axis reversal or diabatic field reversal. Argon cryopumping is then highly desireable. This final design offers almost any option which could be desired for experiments.

PERFORMANCE

The Lamb-shift sources at Giessen,[16] LAMPF,[45] and TRIUMF[46] have all operated with polarized beam intensities from the source ≥1μA. To date the best reported current obtained is the 1.7 μA with P_b=0.70 obtained at Giessen.[47] This source is short, compact, equipped with large pumps for good vacuum and employs the diabatic field-reversal scheme. The source at LAMPF is different in that it contains a spin filter and is longer to allow room for the rapid spin-axis-reversal system now being developed. Like the Giessen source, high pumping speed is provided throughout and is found to be essential. It can be operated to produce $\vec{H}^-$ currents[27] with an intensity of 0.2 μA and P_b= 0.90 or, with higher cesium and argon densities, to produce 1 μA with P_b= 0.60. Clearly with the expected addition of the diabatic field-reversal capability to this source beam currents and polarizations matching the best obtained at Giessen should be available.

CONCLUSIONS

The Lamb-shift source will thrive only to the extent that its $\vec{H}^-$ (or $\vec{D}^-$) output currents can be improved to match those from improving atomic-beam-type sources.[48,49] At the moment, the possibility of space-charge beam blowup and "self-quenching" of the H_0(2S) atoms in space-charge electric fields loom as real barriers preventing substantial improvement in intensity. Clearly _if_ these problems can be overcome by careful design or a clever, new idea,† the versatile spin-state selection, easy spin-axis manipulation, and excellent beam emittance would make the continued use of these sources a certainty. If the output beam intensity does not continue to improve, the future of Lamb-shift sources will be clouded.

REFERENCES

*Work supported in part by the U. S. Department of Energy.

†See the Working Group Summary on Lamb-shift sources in this volume.

1. W. E. Lamb, Jr. and R. C. Retherford, Phys. Rev. 79, 549 (1950).
2. L. Madansky and G. E. Owen, Phys. Rev. Lett. 2, 209 (1959).

3. Bailey L. Donnally and William Sawyer, Physical Review Letters 15, 439 (1965).
4. T. B. Clegg, G. R. Plattner, L. G. Keller, and W. Haeberli, Nucl. Instr. Meth. 57, 167 (1967).
5. R. W. Dunford, R. R. Lewis, and W. L. Williams, Phys. Rev. A13, 2451 (1978).
6. E. G. Adelberger, T. A. Trainor, E. N. Fortson, T. E. Chupp, M. Z. Iqbal, and H. E. Swanson, Nucl. Inst. Meth. 179, 181 (1981).
7. Thomas B. Clegg, Proc. of Symposium on Ion Sources and the Formation of Ion Beams, BNL-50310, edited by Th. Sluyters (Brookhaven National Laboratory, Upton, Long Island, New York, 1971), p. 223.
8. W. Haeberli, Ann. Rev. Nuc. Sci. 17, 373 (1967).
9. Thomas B. Clegg, Proc. of Fourth International Sympsoium on Polarization Phenomena in Nuclear Reactions, eds. W. Grüebler and V. König (Berkhauser Verlag, Basel, 1976), p. 111.
10. W. Haeberli, Proc. of Conf. on High Energy Physics with Polarized Beams and Polarized Targets, ed. by G. H. Thomas (Am. Inst. of Physics, New York, 1977), 51, p. 269.
11. W. Grüebler and P. A. Schmelzbach, Proc. of Conf. on Polarization Phenomena in Nuclear Physics--1980, eds. G. G. Ohlsen et al. (Am. Inst. of Physics, New York, 1981), 69, p. 848.
12. Bailey L. Donnally, Proc. Third Int. Symposium on Polarization Phenomena in Nuclear Reactions, eds. H. H. Barschall and W. Haeberli, (University of Wisconsin Press, Madison, Wisconsin, 1971), p. 295.
13. L. D. Knutson, Phys. Rev. A2, 1878 (1970); H. Bruckman, D. Finken, and L. Friedrich, Phys. Lett. 29B, 223 (1969).
14. P. A. Schmelzbach, W. Grüebler, V. König, and B. Jenny, Proc. of Conf. on Polarization Phenomena in Nuclear Physics--1980, eds. G. G. Ohlsen et al. (Am. Inst. of Physics, New York, 1981), 69, p. 899.
15. P. Schiemenz, D. Ehrlich, R. Frick, and G. Graw, ibid, p. 893.
16. W. Arnold, H. Berg, H. H. Krause, J. Ulbricht, and G. Clausnitzer, Nucl. Instr. Meth. 143, 441 (1977).
17. C. R. Howell and S. A. Wender, submitted to Nucl. Instr. Meth.
18. W. Arnold, H. Berg, and G. Clausnitzer, Proc. of Conf. on Polarization Phenomena in Nucl. Physics--1980, eds. G. G. Ohlsen et al. (Am. Inst. of Physics, New York, 1981), 69, p. 899.
19. T. B. Clegg, S. M. Mitchell, H. L. Manning, and K. Murphy, Triangle Universities Nuclear Laboratory Annual Report--TUNL XVII, ed. by D. R. Tilley (Duke University, Durham, N.C., 1978), p. 86.
20. T. B. Clegg, G. A. Bissinger, and T. A. Trainor, Nucl. Inst. Meth. 120, 445 (1974).
21. P. Schiemenz, private communication.
22. R. A. Hardekopf, Proc. of Conf. on Polarization Phenomena in Nuclear Physics--1980, eds. G. G. Ohlsen, et al. (Am. Inst. of Physics, New York, 1981), 69, p. 896.
23. R. Risler and T. A. Trainor, Univ. of Washington Nuclear Physics Laboratory Annual Report, ed. J. G. Cramer (Univ. of Washington, Seattle, 1978), p. 11.

24. T. A. Trainor and D. W. Holmgren, Bull. Am. Phys. Soc. 26, 603 (1981).
25. S. M. Mitchell, W. R. Wylie, and T. B. Clegg, Triangle Universities Nuclear Laboratory Annual Report--TUNL XVIII, ed. by D. R. Tilley (Duke University, Durham, North Carolina, 1979), p. 107. See also ref. 19.
26. A. S. Schlachter, Proc. of Symp. on the Production and Neutralization of Negative Hydrogen Ions and Beams, BNL-50727, ed. by K. Prelec (Brookhaven National Laboratory, Upton, Long Island, New York, 1977), p. 11.
27. P. Chamberlin, private communication.
28. P. Schmor, private communication.
29. G. Clausnitzer, private communication.
30. J. L. McKibben, G. P. Lawrence, and G. G. Ohlsen, Phys. Rev. Lett. 20, 1180 (1968).
31. W. E. Lamb, Jr. and R. C. Retherford, Phys. Rev. 81, 222 (1951); W. E. Lamb, Jr., Phys. Rev. 85, 259 (1952).
32. J. L. McKibben, Am. Jour. of Phys. 45, 1022 (1977).
33. One can do this by selecting the $m_I = -1$ hyperfine state and ionizing in argon inside a magnetic field of $\sim$9.7G. See ref. 8, p. 383.
34. G. G. Ohlsen, private communication.
35. Y. Tagishi and J. Sanada, Nucl. Instr. Meth. 164, 411 (1979).
36. G. G. Ohlsen, J. L. McKibben, G. P. Lawrence, P. W. Keaton, Jr., and D. D. Armstrong, Phys. Rev. Lett. 27, 599 (1971).
37. P. G. Sona, Energia. Nucleare 14, 295 (1967); T. B. Clegg, G. R. Plattner, and W. Haeberli, Nucl. Instr. Meth. 62, 343 (1968).
38. J. M. Potter, J. D. Bowman, C. F. Hwang, J. L. McKibben, R. E. Mischke, E. E. Nagle, P. G. Debrunner, and L. P. Sorenson, Phys. Rev. Lett. 33, 1307 (1974).
39. E. G. Adelberger, H. E. Swanson, M. D. Cooper, J. W. Tape, and T. A. Trainor, Phys. Rev. Lett. 34, 402 (1975).
40. J. L. McKibben, Proc. of Conf. on Polarization Phenomena in Nuclear Physics--1980, eds. G. G. Ohlsen, et al. (Am. Inst. of Physics, New York, 1981), 69, p. 830.
41. T. A. Trainor, W. B. Ingalls, H. E. Swanson, and E. G. Adelberger, IEEE Transactions on Nucl. Sci. NS-24, 1527 (1977).
42. R. T. Avery, G. R. Lambertson, and C. D. Pike, IEEE Transactions on Nucl. Sci. NS-18, 885 (1971).
42a. Available from Helipot Corporation, So. Pasadena, California.
43. J. G. Cramer, E. Preikschat, and W. Trautman, Univ. of Washington Nuclear Physics Laboratory Annual Report (Univ. of Washington, Seattle, 1977)p. 14; See also ref. 3.
44. J. L. McKibben, G. P. Lawrence, and G. G. Ohlsen, Proc. Third Int. Symp. on Polarization Phenomena in Nuclear Reactions, eds. H. H. Barschall and W. Haeberli (University of Wisconsin Press, Madison, Wisconsin, 1971), p. 828.
45. E. P. Chamberlin, R. L. York, H. E. Williams, and E. L. Rios, Proc. of Conf. on Polarization Phenomena in Nuclear Physics--1980, eds. G. G. Ohlsen et al. (Am. Inst. of Physics, New York, 1981) 69, p. 887.

46. J. L. Beveridge, G. Dutto, P. W. Schmor, and G. Roy, Prof. of Conf. on High Energy Physics with Polarized Beams and Polarized Targets, ed. G. H. Thomas, (Am. Inst. of Physics, New York, 1979), 51, p. 341.
47. G. Clausnitzer, private communication; see also ref. 11.
48. D. Hennies, R. S. Raymond, L. W. Anderson, W. Haeberli, and H. F. Glavish, Phys. Rev. Lett. 40, 1234 (1978).
49. P. A. Schmelzbach, W. Grüebler, V. König, and B. Jenny, Proc. of Internat. Symp. on High Energy Physics with Polarized Beams and Polarized Targets, Lausanne, 1980, ed. C. Joseph and J. Soffer, (Birkhauser Verlag, Basel, 1981), p. 432.

WORKING GROUP SUMMARY - POTENTIAL IMPROVEMENTS IN LAMB-SHIFT POLARIZED ION SOURCES

Thomas B. Clegg
Department of Physics, University of North Carolina*
Chapel Hill, North Carolina 27514
and
Triangle Universities Nuclear Laboratory, Duke Station*
Durham, North Carolina 27706

INTRODUCTION

The working group opinion about the potential for improving Lamb-shift polarized sources has changed from pessimistic to optimistic as we have looked hard at the problems confronting us. The most attractive feature of the Lamb-shift source is that it employs reliable, mature technology which has been proven in a number of laboratories. It has the possibility for versatile spin-state selection and rapid spin-axis reversal which facilitate extremely difficult measurements. The basic question about the Lamb-shift source is whether the output $\vec{H}^-$ beam intensity can be increased substantially beyond the $\sim$1 μA levels attained in several laboratories for several years. Substantial effort has gone into trying to increase this intensity, but progress has slowed.

The basic question is whether the useable metastable 2S atomic beam intensity can be increased within the source. The low 550 eV H^+ (or 1100 eV D^+) beam energy means there are severe space-charge-related beam-divergence problems. Space charge also creates internal electric fields within the beam which can be large enough to quench the 2S atoms to the 1S ground state via Stark mixing of the 2S and 2P levels. Thus, reducing space charge within the beam is <u>the</u> central problem to be overcome in improving polarized beam intensity.

Another advance which would be a convenience to experimentalists would be the development of a rapid-spin-reversal technique for Lamb-shift sources utilizing the Sona scheme for obtaining nuclear polarization. Thus far the only fast-flip processes available require first that a single hyperfine state be selected in a spin-filter, reducing the output beam intensity for $\vec{H}^-$ by roughly a factor of two from that which is available from sources which employ the Sona scheme. Substantial beam current increases from <u>any</u> improvements would, however, mitigate this loss. Nevertheless, a clever idea to merge the Sona scheme with rapid-spin-axis reversal is needed.

In the sections below these problems and possible solutions discussed within the working groups are investigated further.

*Work supported in part by the U. S. Department of Energy.

METHODS TO IMPROVE METASTABLE BEAM INTENSITY

Microscopic Quenching

One type of quenching which one can imagine for the 2S metastable atoms arises because of microscopic electric fields between individual ions and electrons in the beam. An attempt was made to estimate the magnitude of these E-fields to determine if they would be a problem.

It was assumed that 500 μA of H^+ beam passes through the cesium canal in the best, present Lamb-shift sources. Of this, only $\sim$100 μA is useful beam which can be changed to 1.5 μA of $\vec{H}^-$ and accelerated. Thus there is substantial beam loss already within existing sources. One finds that

$$I = 500\ \mu A = 3.1 \times 10^{15}\ \frac{\text{ions}}{\text{sec}}\ .$$

If the cesium canal is 1 cm^2 in cross sectional area, this same number constitutes the flux density passing through the canal. For 550 eV H^+, the velocity is 3.1×10^7 cm/sec, so the spatial density of ions within the canal is

$$\rho = \frac{I}{v} = 10^8\ \frac{\text{ions}}{\text{cm}^3}\ .$$

From this one can calculate

$$\langle r \rangle = \text{Average spacing between ions}$$
$$= \left(\frac{1}{10^8}\right)^{1/3} = 2.1 \times 10^{-3}\ \text{cm} \cong 10^{-3}\ \text{cm}$$

Realize that the current would have to increase a factor of eight over that assumed here to make the individual charges be spaced closer than 10^{-3} cm. This should account for an axially peaked beam profile within the canal and the presence of Cs^+ ions and electrons there. The electric field at the site of an H_o(2S) atom at a distance $\langle r \rangle$ from an H^+ ion is thus

$$E = \frac{q}{4\pi\varepsilon_o \langle r^2 \rangle} = 0.14\ \frac{\text{volts}}{\text{cm}}$$

Since this field should be $\gtrsim$ 10 v/cm before quenching of 2S atoms becomes significant, at this H^+ current level microscopic quenching should not be a problem. Indeed even if the H^+ current through the canal were to increase by the unattainable factor of 10^3, one would find $E_{microscopic}$ = 14 v/cm, a magnitude which is still probably manageable. Thus it seems one does not need to worry about this effect.

Macroscopic Quenching

The more familiar electric fields which might result in 2S-atom quenching result from averaging contributions from all charges within the beam. This produces a radial electric field at any radius r which can be calculated to be

$$E = \frac{I}{2\pi\varepsilon_o r v}$$

where I is the current in the beam inside radius r and v is the velocity of ions in the beam. Assuming again I = 500 μA and assuming pessimistically that the beam has a dense core with R = 3 mm containing most of the ions, one calculates that at this radius

$$E \cong 100 \frac{\text{volts}}{\text{cm}}$$

Quenching 2S atoms in this field can clearly be a problem. However, observation of the present $\vec{H}^-$ output currents probably implies that these macroscopic fields are substantially reduced by slow electrons trapped within the beam. How well this is accomplished now is not known. In several sources utilizing duoplasmatrons with the close-coupled geometry, experimenters observe an increase in the $\vec{H}^-$ beam intensity when electric deflection fields are applied immediately following the cesium canal. This effect is interpreted as meaning that rapid removal of the ions from the 2S-atomic beam reduces this macroscopic quenching. If this is true, clearly magnetic rather than electric deflection would be more effective. The macroscopic field could be reduced by accepting the same current I within a larger beam diameter in the cesium canal, but this is likely to make more difficult the separation of $\vec{H}^-$ ions from unwanted, unpolarized H^- ions emerging from the cesium canal.

Complete space-charge neutralization of the beam is a necessary goal, but this will be difficult in the dynamic environment of the cesium canal. An entering H^+ current I emerges with 5% of the beam as H^- ions and the rest roughly equally divided between H^+ ions, 2S metastable atoms, and 1S ground state atoms.[1] Thus the number of electrons required for neutralizing the H^+ ion space charge will vary along the axis through the cesium canal.

Definitive measurements to reveal the presence or absence of this macroscopic quenching effect are badly needed. The Lyman-α detector, used at TUNL to monitor metastable beam intensities and profiles,[2] should be capable of detecting this quenching as a function of beam radius if the light entering it is tightly collimated and if the detector is mounted on a platform precisely movable to scan the beam radially and axially. These measurements will be attempted by P. Chamberlin on the LAMPF source in the near future. We anxiously await their outcome.

Potential Improvement Using an ECR Source

At Grenoble Geller et al.[3-5] have developed a source based on the electron cyclotron resonance (ECR), which has many features which would appear to make it ideal for producing H^+ and D^+ ion beams for a Lamb-shift source. This source has produced D^+ ion beams at 1 keV and transported them through a cesium charge-exchange region exactly as required for the Lamb-shift source, with a D^+ current density of 60 mA/cm², an energy spread of the ions in the beam of 1 to 2 eV, an operating pressure of $\sim10^{-4}$ Torr, and 80% of the output beam as D^+ ions. These properties are accomplished with a 4kG axial magnetic field covering the entire ion source and cesium charge-exchange region. The presence of this field enhances trapping of slow electrons and accomplishes almost complete (>99%) space-charge neutralization. The divergence of the neutral beam emerging from the cesium canal has been measured to be equal to that of the ions as they are extracted from the ion source, $\sim4^\circ$ half angle. Thus a 1 cm diameter $D_o(2S)$ beam emerging from the cesium canal would be only ~3 cm diameter after drifting 1.5 m, the maximum distance which would be required in an operational Lamb-shift source.

Projections of what $\vec{D}^-$ current one might obtain from such a source by simply using these numbers and the known charge-exchange efficiencies in cesium and argon yield astonishingly large currents, $\sim$900 μA! If the space-charge neutralization is as good as is claimed, then one might assume the macroscopic electric fields would not be much larger than those present in existing Lamb-shift sources. Thus, one might be extremely optimistic. Nevertheless, there are two problems which seem potentially serious enough that they diminish considerably our hope for the projected $\vec{H}^-$ current.

First, the beam must emerge from the magnetic field following the cesium canal into a 575G region for spin-filter or Sona-scheme operation. Ordinarily a decrease in axial field of $\sim$100G/cm is possible without quenching metastable 2S atoms in the desired m_J = +1/2 hyperfine states. This quenching occurs in $\vec{v} \times \vec{B}_r$ motional electric fields in this fringing field region where $B_r=-(r/2)(dB_z/dz)$. In this case, however, emerging from the 4kG axial field takes the beam through the α-d level crossing[6] near 2.3kG where atoms in the α-state have their lifetime shortened so much that they almost surely would be lost by quenching, even for very small electric fields. Thus any ECR source used on a Lamb-shift polarized source must operate inside a confining magnetic field substantially less than 2.3 kG. We do not know at present how much this might decrease the useable H^+ beam current.

The second potential serious drawback is the difficulty of maintaining space-charge neutralization within the beam in the fringe-field region after the cesium canal. Any reduction in the nearly complete compensation inside the Grenoble ECR source will certainly lead to macroscopic quenching fields. We cannot predict now how bad these might be.

Implementation of the ECR source on a Lamb-shift source is not inexpensive. The 200 to 300 watt, 10 GHz amplifier used at Grenoble costs ∿$20000, and the magnet coils and their power supplies add significant further expense. Such a system will be assembled at the Univ. of Alberta within the next six months and tests will begin to see if beam intensities can be enhanced from the Lamb-shift polarized source at TRIUMF.[7] There is real optimism that some beam current increase will result.

FAST SPIN-AXIS-REVERSAL AND THE SONA SCHEME

During the discussion of optical pumping developments for ion sources, D. E. Murnick pointed out that if the cesium canal in a Lamb-shift source could be optically pumped so the cesium atoms were all in a single "pure" or "stretched" hyperfine state, so the spin configuration is independent of magnetic field B, then one could pick up a polarized electron to form electron-polarized $H_o(2S)$ atoms. In the case of the electron-pickup in a polarized alkali vapor to try to form electron-polarized $H_o(1S)$, one has real problems with pickup into higher excited states and loss of polarization when these excited-state atoms decay. Here, however, pickup into the $H_o(2S)$ states is strongly dominant over pickup to higher states. Thus, if suitable cesium optical pumping can be achieved at the normal densities of $3x10^{13}$ atoms/cm^3 required,[8] one has real potential for improving the Lamb-shift source.

Assuming that the optically pumped cesium oven were possible by shining circularly polarized laser light along an axis colinear with the beam, then all $H_o(2S)$ atoms formed would be in the α-states (or β-states) depending on the handedness of the circular polarization. This immediately doubles the output $\vec{H}^-$ beam intensity because it eliminates the need for quenching half the $H_o(2S)$ atoms which are in unwanted hyperfine states. Furthermore, it eliminates the need for the first 575G coil in the Sona scheme and facilitates a physically shorter ion source with perhaps slightly higher resulting intensity. Last and most importantly the spin-quantization axis can be flipped rapidly and easily by flipping the handedness of the laser light.

REFERENCES

1. A. S. Schlachter, Proc. of the Symposium on Production and Neutralization of Negative Hydrogen Ions and Beams, BNL 50727, edited by K. Prelec (Brookhaven National Laboratory Upton, Long Island, New York, 1978) p. 11.

2. S. M. Mitchell and T. B. Clegg, Triangle Universities Nuclear Laboratory Annual Report - TUNL XVII, edited by D. R. Tilley (Duke University, Durham, N. C., 1978) p. 86.

3. R. Geller, C. Jacquot and P. Sermet, Proc. of Second Symposium on Ion Sources and The Formation of Ion Beams, LBL 3399, ed. by C. P. Pezzotti (Lawrence Berkeley Laboratory, 1974) p. III-5.

4. R. Geller, <u>et al</u>., Proc. of The Symposium on The Production and Neutralization of Negative Hydrogen Ions and Beams, BNL 50727, edited by K. Prelec (Brookhaven National Laboratory, Upton, Long Island, New York, 1977) p. 173.

5. M. Delaunay <u>et al</u>., Proc. of The Second International Symposium on The Production and Neutralization of Negative Hydrogen Ions and Beams, BNL 51304, edited by Th. Sluyters (Brookhaven National Laboratory, Upton, Long Island, New York, 1980) p. 255.

6. W. E. Lamb, Jr. and R. C. Retherford, Phys. Rev. <u>79</u>, 549 (1950).

7. P. Schmor, private communication.

8. N. D. Bhaskar, J. Pietras, J. Camparo, W. Happer, and J. Livran, Physical Review Letters <u>44</u>, 930 (1980).

Report on the Lamb-shift PPIS Work Session I

Wednesday, May 20, 1981

Chairman: T.B. Clegg Secretary: M.A. Cummings

This session centered exclusively on discussion of the metastable beam in Lamb-shift polarized sources. There was a general agreement among participants to focus on the fundamental limitations of the intensity of $H_0(2S)$ atomic beams. Consideration of technical problems of the various polarized source systems followed.

Another question was how large the beam could be upon entering the cesium canal. It was believed to be of some benefit to have a wide beam if it was of small emittance. There was a discussion on whether multi-aperture sources of H^+ ions would be more effective for filling a cesium canal of a given diameter than present H^+ sources. A quick calculation extrapolated from Berkeley results at 35 keV yielded 4 mA maximum current into the cesium canal at 1 keV, indicating that they might improve things. However a more careful calculation should be made to check this result.

P. Schiemenz described a calculation which assumed a uniform parallel beam, no space charge neutralization, and no magnetic field. He estimated that from a 1 cm diameter beam he could obtain 1 μA of polarized H^-, and from a 3 cm diameter beam, 3 to 4 μA of polarized H^- beam. G. Clausnitzer said that for measurements with cesium canal diameters of 1, 1.25 and 1.5 cm he obtained 1, 1.25 and 1.5 μA of polarized H^- beam respectively, with the corresponding beam polarization values ranging from 0.70 to 0.67. The polarized H^- beam intensity increases linearly with the canal diameter and not as the square of that diameter.

Schiemenz's calculations implied a loss of metastable atoms for larger radii from quenching by space charge. However, Clausnitzer's interpretation of his measurements was that the linear dependence on the diameter resulted from the non-uniform radial illumination of the cesium canal. The measurements, therefore, could neither prove nor disprove that there is a significant loss of $H_0(2S)$ intensity arising from "self-quenching" of the $H_0(2S)$ atoms by the macroscopic electric fields of the beam. Clausnitzer made the point that he felt both microscopic and macroscopic quenching effects were important in the beam. All present believed that the macroscopic fields were certainly significant at some level. The question to be dealt with is whether it is possible to estimate the size of the microscopic contributions by estimating the internal fields from densities of the various species of ions. An accurate calculation would be quite difficult, however.

Clausnitzer expressed doubt that separation of the unpolarized H^- beam from the cesium canal and the polarized H^- ions coming from the argon charge-exchange would be easy, or even possible, for large beam diameters. Schiemenz said that his calculations show no significant difference when comparing a small diameter, diverging beam such as at the Triangle Universities Nuclear Lab and a large diameter, converging beam such as in his source at Munich.

Clausnitzer said furthermore that 100μA is the Child-Langmuir limit for a beam which can be extracted from his source and put into the cesium canal. He obtains, however, 500 μA experimentally. This, he said, is proof that even in the "extended geometry" sources, significant space charge neutralization occurs.

From the above discussions the group acknowledged the fact that despite the variety of systems used, everyone seemed to be reaching the same beam intensity limit. The intensities now obtained required much work and the gut feeling of all present was that at best perhaps a factor of 2 in $H_0(2S)$ intensity is possible. At present, however, what is clearly not understood is which factors are the most important limitations.

Report on the Lamb Shift PPIS Work Session II

Thursday, May 21, 1981

Chairman: T.B. Clegg Secretary: M.A. Cummings

The group continued to concentrate on finding ways to improve the overall $H_0(2S)$ beam intensity. The main discussions were as follows:

1. The Berkeley multi-aperture source was discussed by R. Mobley. He made an effort to extrapolate the performance for 40 keV ions to 1 keV H^+ ions. His basic source is as shown. From these results of previous work, at 40 keV, assuming a 1 cm diameter hole from which to extract H^+ beam, he calculated first that $\pi r^2 \times 500$ mA/cm^2 $\cong$ 390 mA would be extracted. Using the Child-Langmuir law,

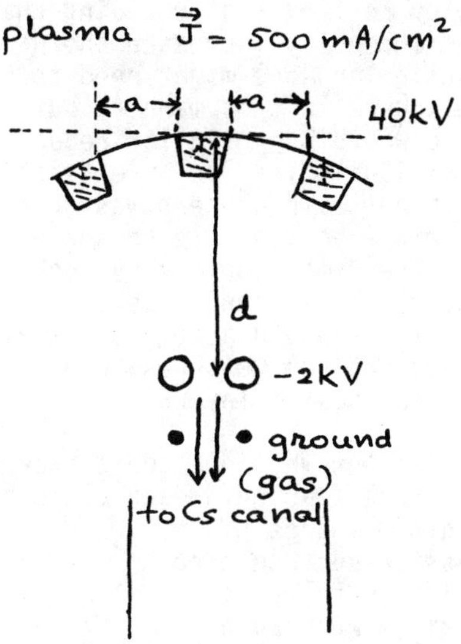

$$I = k \frac{(\text{Area})\, V^{3/2}}{d^2}$$

, he extrapolated it to 390 × (1 keV/40 keV)$^{3/2}$ $\cong$ 1.5 mA of extracted current at 1 keV. This is not too different from present experience with the duoplasmatrons now used in Lamb-shift sources.

Then he proposed that the dimensions of the source could be scaled down, while maintaining the same aspect ratio $\pi(a/2)^2/d^2$, to make the same current available from a smaller aperture. Then one can make a "pepper shaker" array of these smaller apertures to fill the original 1 cm diameter. The available current would then increase directly as n, the number of holes. The overall emittance of such a pattern would ideally not be larger than that of the original source, as seen in the emittance diagram at the right.

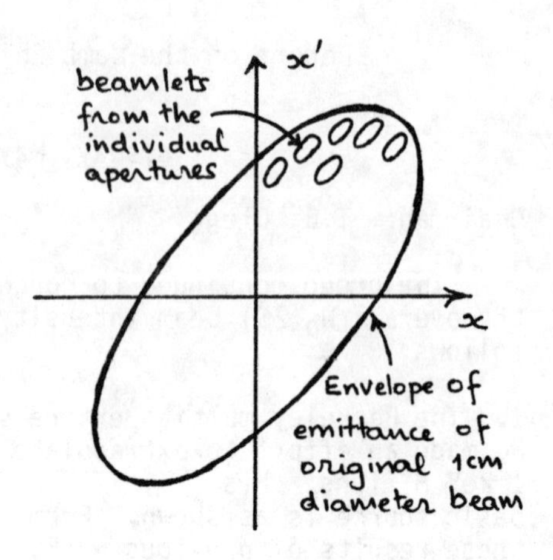

Discussion developed and concern was expressed that the gas flow from such a multi-aperture source would be so large that metastable atoms produced would be quenched by collisions. Certainly much higher pumping speeds and perhaps differential pumping would be needed. Further concern was expressed that for the Berkeley source, H^+/H_2^+ ratio was small, maybe ~ 50%, whereas some sources operate with ~ 80% presently. Mobley replied that reducing the plasma density increased the H^+ fraction and one might need to run the source in this manner, but there would be a similar reduction in available ions to be extracted. The other main concern was that one might not be able to scale down the dimension d very much because the distances between electrodes would become too small to support the voltages, especially in the presence of the high gas flow and cesium layers.

As for what one could expect, if the pole dimension a and electrode spacing d each could be shrunk a factor of 4 and then the original source emission-surface area filled as fully as possible with such holes, one should be able to extract approximately 12 times the original current of 1.5 mA ≅ 18 mA.

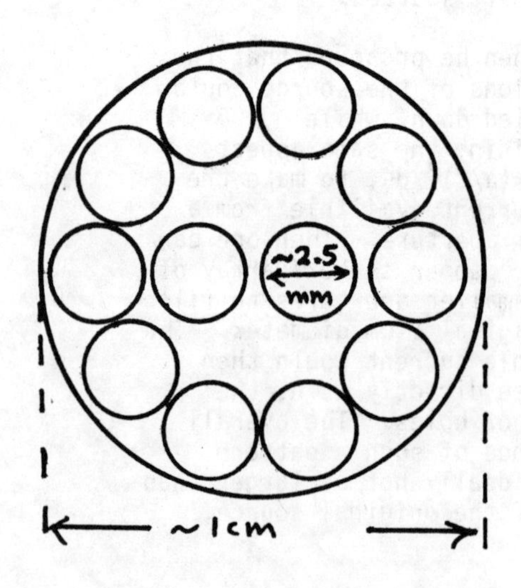

Space charge effects will clearly be very important. In the geometry close-coupled to the cesium canal, one still might hope this could be minimized. One might need to retune considerably the plasma parameters of the source to shape the "bubbles" of the plasma surface in each emission aperture. It is not clear in the highly interdependent parameter space of the source and cesium canal, how easy it would be to optimize the beam. Still the potential for somewhat larger current is present if one is willing to accept a larger cesium cell diameter and the accompanying larger overall polarized beam emittance. Hopefully for cesium canals of ~ 2 cm diameter, one might get a significant beam increase without significant increase in the $H_0(2S)$ quenching.

2. Clausnitzer questioned again whether it would be possible to accomodate a larger diameter beam emerging from the cesium canal. He cautioned that it would be very hard to removed unwanted H^-ions emerging from such a larger canal.

3. Clausnitzer reported that, while making his measurements which showed a strong beam intensity dependence on the ratio D/L of the duoplasmatron anode aperture (See Santa Fe proceedings, pg. 882), he found also that there was a minimum of the a.c. ripple ("hash") on the arc voltage when (for D=0.4 mm and L≅0.15 mm) he obtained the maximum polarized H^- output current. No one really understood what this meant, except that there might be internal "plasma oscillations" inside the duoplasmatron arc chamber. These are in the MHz region, and it is not at all obvious why these should depend on the anode-aperture channel length. Clegg reported old measurements at Los Alamos in which McKibben observed the smallest energy spread in the output beam for the same minimum in this a.c. "hash" on the arc supply.

Chamberlin questioned whether there might be significant $H_0(2S)$ quenching from Cs^+ ions in the beam. He reported that he had not observed this when looking at the Lyman-α detector signal when his beam was pulsed on. Presumably if this were a problem, the output polarized current would sag after an initial peak as more Cs^+ ions were created by the beam during one pulse. He observed good square-top pulses.

Report on the Lamb-shift PPIS Work Session III

Monday, May 25, 1981

Chairman: T.B. Clegg Secretary: M.A. Cummings

Discussion concentrated on what definitive measurements could be made to improve Lamb-shift source operation and to learn what features were real limitations. The following suggestions were made.

1. The Lyman-α detectors now available in laboratories at TUNL, Seattle, and Los Alamos should be used behind tight collimation to study the radial production of 1216Å radiation. The goal would be to determine where the metastables are being quenched within the beam profile. Measurements before and after the cesium canal, and along the beam toward the spin filter should help us learn how serious "self-quenching" effects are within the beam. Once such a system is assembled, perhaps collisional quenching rates could also be studied by bleeding in small amounts of various gases. After further discussion, P. Chamberlin said he would try to make such measurements at LAMPF.

2. Magnetic fields instead of electric fields should be used after the cesium canal to deflect charged ions more effectively.

3. P. Chamberlin suggested also that one might learn something about space-charge forces by putting a thin wire axially down the axis of the cesium canal and applying a bias voltage to it. This can not cancel macroscopic space charge fields exactly since this field falls logarithmically with r and the space-charge increases linearly with r if the beam intensity profile is uniform.

4. The $\vec{H}$-beam always increases as the vacuum improves. Surfaces coated with Zeolite and kept at 20°K can be used near the beam to lower the pressure locally.

5. Discussion continued on finding inexpensive ways to incorporate the ECR source into a Lamb-shift source. P. Schmor suggested that the Klystron from a microwave oven might be a cheap source of r.f. power. P. Schiemenz said that he would try to install a pair of magnets over his r.f. bottle on his source to try to make a better electron trap in the plasma and enhance the beam output.

Report on the Lamb-shift PPIS Work Session IV

Tuesday, May 26

Chairman: P. Schiemenz Secretary: M.A. Cummings

The emphasis of this session was on improving existing Lamb shift sources rather than on new schemes.

It was generally agreed that a good vacuum (10^{-6} torr or better) in the region of H(2S) and downstream was essential, but was not so important in the H^+ region.

One idea brought up was to tilt the positive ion source at a slight angle (5-10°) from the rest of the source apparatus, and then to deflect the H^+ beam into the Cs canal with a dipole magnet. This would automatically remove H_2 and H_3 ions and would reduce the unpolarized background which arises from neutrals emerging from the positive ion source and being charge exchanged in the argon canal. There was evidence from a measurement at Los Alamos that the unpolarized background can be reduced considerably this way. Magnetic deflection of the H^+ and H^- ions after the Cs canal was discussed. It should be possible to completely deflect the charged particles out of the beam before entering the polarizer.

It was pointed out again that a Lyman-α detector which can be moved from the Cs canal to the argon canal to measure the meta-stables would help to learn how much self-quenching occurs. Another concern was the amount of argon present in the region where selection of the proper hyperfine state is made. To find this out one would reduce pumping in this region, measure the polarization and extrapolate the results to those at a better vacuum.

It may be advantageous to have a magnetic field ~ 50 gauss between the positive ion source and the Cs canal in extended geometries to improve the space charge neutralization. One would also like to have some method to measure the diameter of the beam, and to check on improvement (perhaps moving a wire through the beam). The Gabor lens was suggested as a type of lens which could preserve space charge neutralization.

There was some discussion of the fast spin flip apparatus which is used for parity violating ion experiments. Experience at LAMPF has shown that the vacuum in this device should be considerably improved.

REVIEW OF THE GROUND STATE ATOMIC BEAM STAGE OF POLARIZED HYDROGEN ION SOURCES

W. Grüebler
Laboratorium für Kernphysik, Eidgenössische Technische Hochschule, 8093 Zürich, Switzerland

ABSTRACT

This paper reviews the basic principle of the atomic beam stage for polarized hydrogen ion sources. The problems of the forming, the spin states separation, and the focussing of neutral atomic hydrogen beams are discussed. It is shown that substantial improvement of the intensity and the density of the atomic beam can be obtained by a well-designed multi-stage-separation magnet system, and by cooling the atoms to a low velocity.

INTRODUCTION

The status of conventional polarized ion sources has periodically been reviewed over the past twenty years[1-4]. The two most recent surveys of this topic were given in 1980 at the Santa Fe Conference[5] and the conference of High Energy Physics with polarized particles at Lausanne[6]. While for cyclotron accelerators mostly positive hydrogen ions are required, tandem accelerators demand negative ions for the injection. Since the proposition of Donnally[7] to produce polarized negative ions by the Lamb-shift method, most tandem laboratories equipped with polarized ion sources used this scheme because it was thought that higher beam intensities could be obtained. However, recently the conventional method using a ground state atomic beam attracted considerable attention since large improvements could be achieved either by ionizing the neutral atomic beam directly to H^- by a neutral Cs beam[8] or by ionizing the $\vec{H}^o$ beam by electron bombarding in new powerful ionizers[9,10] and producing H^- by double charge exchange in Na[11,12,13]. While more than 100 μA beam output of polarized positive ions have been found[10,14], for both ionizing schemes about 3 μA H^- ions were reported[8,10]. The basic scheme of ground state atomic beam sources using a thermal atomic beam and their different ionizing modes is shown in fig. 1.

The thermal atomic beams have been in use since the beginning of polarized ion source work, but recently fast, polarized, ground state atomic beams have been produced by charge and polarization pick-up[15]. The scheme of this type of source is shown in fig. 2. Since the problems arising in this source type are discussed thoroughly in the optical pumping sessions of this workshop, I will limit this review to the problems and improvements of thermal atomic beams.

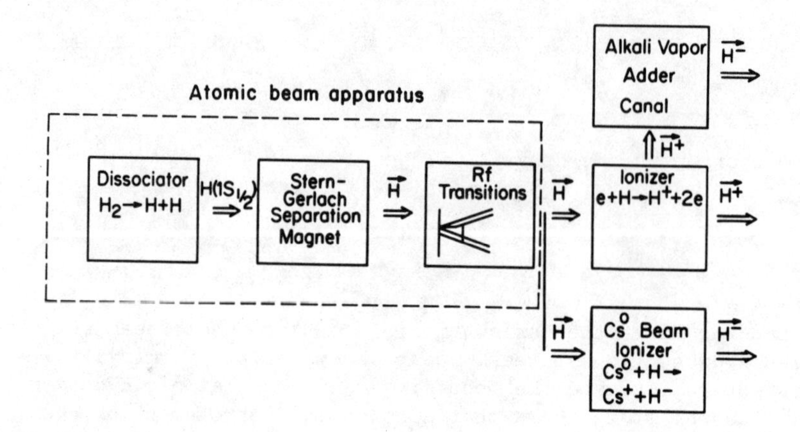

Fig. 1. Schematic diagram of the ground state atomic beam sources with different ionizer schemes.

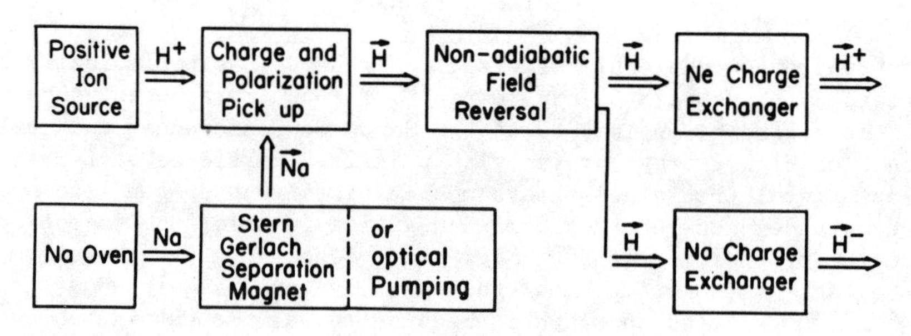

Fig. 2. Schematic diagram for the production of polarized ions by polarization exchange.

The requirements for such an atomic beam apparatus are the following:

1. *High polarization.* High electronic polarization is achieved by Stern-Gerlach separation and high nuclear polarization is produced by rf transitions.
2. *Intensity.* It is interesting to notice that the atomic beam intensities reported over the past 20 years seem to be nearly constant; namely:

   1960: $1 \cdot 10^{16}$ atoms/sec
   1970: $1.6 \cdot 10^{16}$ atoms/sec
   1980: $3 \cdot 10^{16}$ atoms/sec.

   Part of this observation is connected with the difficulties of measuring the neutral beam intensity; in the early investigations the intensities have been largly overestimated but nevertheless, the progress has been quite slow and it is here where large improvements can be expected.

3. *Density*. Since the ion beam intensity I_p is proportional to the density of the atomic beam this quantity is more important than the beam intensity.

4. *Velocity of the atoms*. Since the density is connected with the velocity v by the relation

$$\rho \propto \frac{1}{v} \propto \frac{1}{\sqrt{T}} \quad (1)$$

low velocity is required for a high density. Further a *sharp velocity distribution* is of great advantage for the Stern-Gerlach separation and focussing.

5. *Matching of the beam properties in the ionizer*. It is obvious, that the beam parameters should be matched to the ionizing electron- or Cs^o-beams in the ionizing region as well as possible.

6. *Change of the sign of the polarization*. This feature, which is important for any polarization experiment, is particularly interesting if high precision is required. The reversal of the sign of the polarization should occur without intensity modulation and change of the optical properties of the polarized beam.

SEPARATION OF THE m_j = +1/2 AND m_j = -1/2 STATES

This is a reminder of the method to produce an electron polarized hydrogen beam. Fig. 3 shows the Rabi diagram where the relative energy W/ΔW in units of the hyperfine splitting ΔW is plotted versus the magnetic field χ in units of the critical field of hydrogen atoms B_{crit}. In an inhomogeneous magnetic field the atoms with m_j = +1/2 are separated

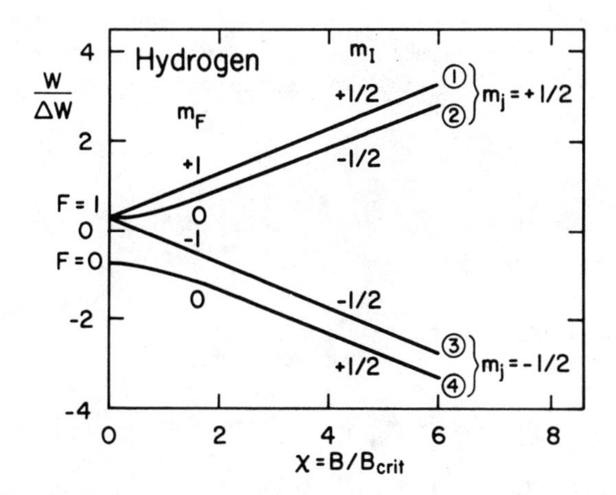

Fig.3. Energy-level diagram of the hydrogen atom in a magnetic field (Rabi diagram).

from $m_j = -1/2$, since a force

$$F = \mu_{eff} \cdot \text{grad } B$$

is acting on the hydrogen atoms. The quantity μ_{eff} is the effective magnetic moment of the particle. In atomic beam sources mostly magnetic sextupole fields are used for this Stern-Gerlach separation. Such a field configuration is shown in fig. 4, r_o being the pole tip radius

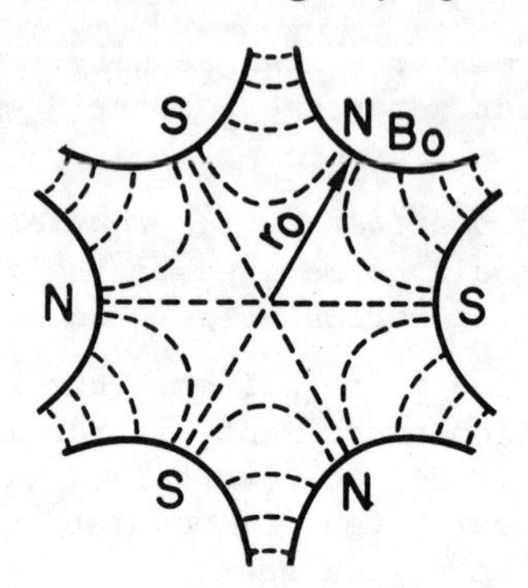

Fig. 4. Field configuration of a sextupole separation magnet.

and B_o the pole tip field. The radial field distribution is given by the relation

$$|\vec{B}(r)| = B_o(\frac{\vec{r}}{r_o})^2 \tag{2}$$

and the field gradient by

$$\text{grad } |\vec{B}(r)| = \frac{2B_o}{r_o^2} \cdot \vec{r} \tag{3}$$

In a sextupole magnet with a strong field and a constant r_o the trajectories of atoms entering the magnet on a radius r_a with an angle α can be calculated by solving the differential equation

$$mv^2 \frac{d^2r}{dz^2} = -\frac{2\mu_B \cdot B_o}{r_o^2} \cdot r \tag{4}$$

where μ_B is the Bohr magneton and z the symmetry axis of the magnet. The states 1 and 2 with $m_j = +1/2$ have $\mu_{eff} = -\mu_B$. The solution of equation (4) is

$$r(z) = A_1 \cos\xi \cdot z + A_2 \sin\xi \cdot z \tag{5}$$

with

$$\xi^2 = \frac{2\mu_B \cdot B_o}{mv^2 \cdot r_o^2} \quad \text{and} \quad A_1 = r_a \; ; \; A_2 = \frac{\alpha}{\xi} \quad . \tag{5a}$$

The trajectories calculated for $T = 295^oK$, $B_o = 1.0\,T$, entrance radii $r_a = 0, 1, 2$ mm, and angles $\alpha = 0^o, \pm1^o, \pm2^o, \pm3^o$ are shown in fig. 5. The solution for the states 3 and 4 with $m_j = -1/2$ and $\mu_{eff} = +\mu_B$ is

$$r(z) = A_1 \text{Cosh}\xi \cdot z + A_2 \text{Sinh}\xi \cdot z \tag{6}$$

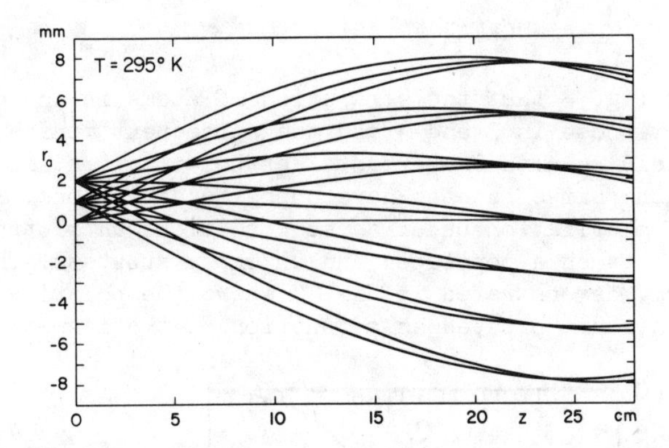

Fig. 5. Calculated trajectories for m_j = +1/2 states in a sextupole magnet for T = 295°K, B_o = 1.0 T, r_o = 8 mm and r_a = 0, 1, 2 mm and α = 0°, 1°, 2°, 3°.

Atoms entering the magnet on the axis are not deflected out of the atomic beam, whereas particles with large angles and radii are separated completely from the atoms in the m_j = +1/2 states. Most critical are atoms entering with small angles with respect to the magnetic axis. A few such trajectories are shown in fig. 6 for T = 295°K, B_o = 1.0 T, r_a = 0, 1, 2 mm and α = 0°, ±1°.

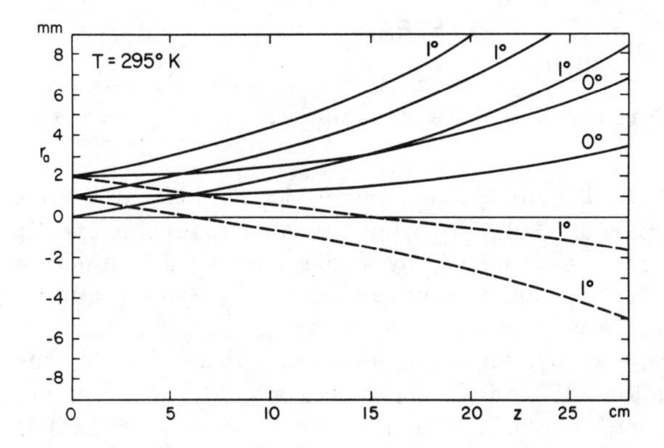

Fig. 6. Calculated trajectories for the m_j = -1/2 states in a sextupole magnet for T = 295°K, B_o = 1.0T, r_o = 8 mm and r_a = 0, 1, 2 mm and α = 0° and 1°.

It is obvious from fig. 3 that the separation of atoms in the electronic states 1 and 2 from those in 3 and 4 and the subsequent ionization in a strong magnetic field results in zero nuclear polarization since the proton spins in each pair of states have opposite directions. In order to obtain nuclear polarization under these conditions an rf transition has to be induced between a populated and an empty substate. The situation is most clearly demonstrated in fig. 7 where the polarizations of the single components are plotted as a function of the magnetic field

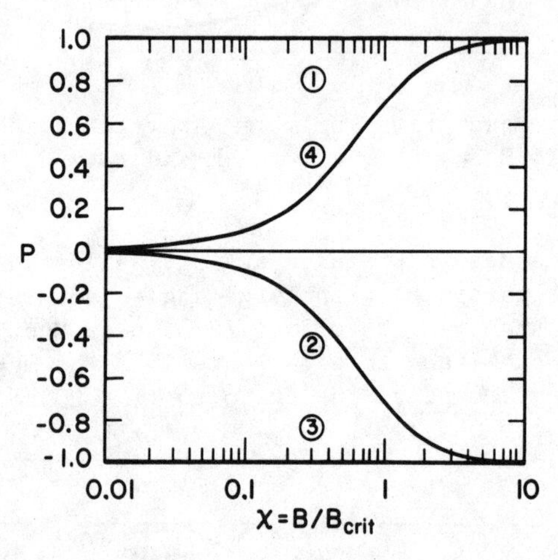

Fig. 7. Polarization P_z of protons in the hydrogen atoms as a function of external field for the 4 single substates.

in the ionizer. A 2 to 4 transition, which is carried out in a relatively strong field compared to B_{crit}, induces a complete proton polarization. The same aim can be reached by the multilevel transition 1 to 3 in a weak magnetic field, but the polarization is inverted. While for both of these transitions a strong ionizing field is required, only a reduced polarization can be achieved without transition by the use of a weak field ionizer. The different combinations of ionizing fields, rf transitions, additional Stern-Gerlach separation and their polarization are summarized in table 1. The maximum polarization can easily be verified with the aid of fig. 7.

Table 1

Ionizer B-field	Transitions	Substates	P_{max}
weak	none	1 + 2	+ 0.5
weak	1 to 3	2 + 3	- 0.5
strong	2 to 4	1 + 4	+ 1.0
strong	1 to 3	2 + 3	- 1.0
any	2 to 4 +Stern-Gerlach separation	1	+ 1.0
any	2 to 4 +Stern-Gerlach separation + 1 to 3	3	- 1.0

OPTICAL PROPERTY OF A SEXTUPOLE MAGNET

A sextupole magnet acts on hydrogen atoms with T = constant and in states 1 and 2 as a thick focussing lens. This property is clearly shown by the trajectories in fig. 5 for T = 295°K. As the temperature (and the velocity of the atoms) decreases the focal length decreases proportionally to the velocity v. This is shown in figs. 8 and 9 by the trajectories of atoms with temperatures of 77°K and 20°K.

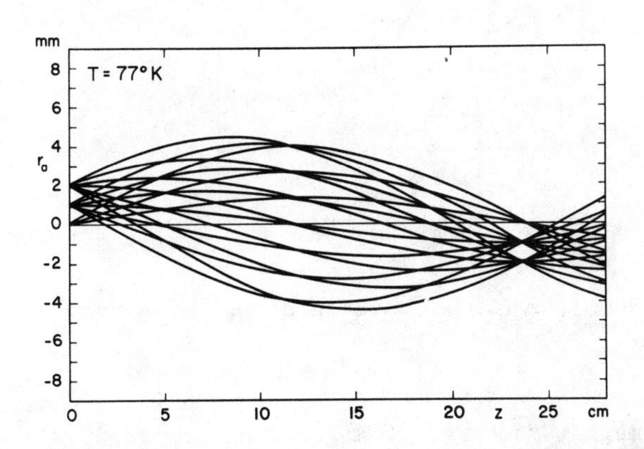

Fig. 8. Calculated trajectories for the m_j = +1/2 states for T = 77°K. For details see fig. 5.

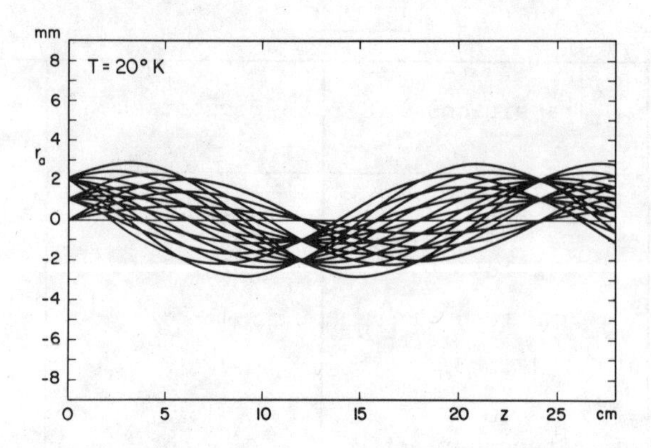

Fig. 9. Calculated trajectories for the $m_j = +1/2$ states for T = 20°K. For details see fig. 5.

The particles in an atomic beam formed by a nozzle at the end of a low pressure rf dissociator have a Maxwell velocity distribution. For this reason the focal length for such a beam entering in a sextupole magnet changes considerably and no longer produces a well focussed beam with an image of the entrance aperture in one single plane. This is demonstrated in fig. 10 where the velocity distribution is superimposed with a diagram of the distance of the focal plane z_f from the entrance as a function of velocity.

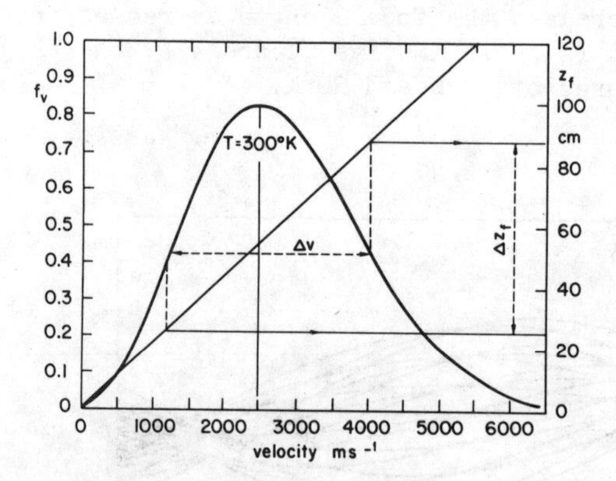

Fig. 10. Maxwell velocity distribution and the corresponding distance z_f to the focal plane.

This situation however can be improved by achromatic focussing using a second sextupole magnet (compressor magnet) as proposed by Glavish[16] some years ago. In this scheme the first sextupole magnet is followed after a drift space, where rf transitions can be located, by a second short "compressor" sextupole magnet, as shown in fig. 11. Slow atoms

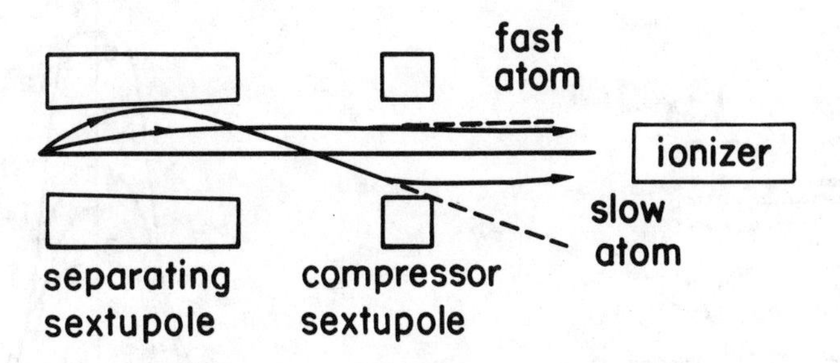

Fig. 11. Schematic illustration of achromatic focussing.

emerging from the first sextupole diverge rapidly due to the focussing properties as shown in fig. 10 and enter the compressor sextupole at a large radius. As they are slow they spend more time in this magnet and are bent strongly towards the axis. On the other hand, fast atoms accepted in the first sextupole are less divergent, spend less time in the compressor and are on a smaller radius. Therefore they are not focussed so strongly by the compressor magnet. By this achromatic focussing the atomic beam density at the ionizer is greatly enhanced. The calculated trajectories of different v for the sextupole doublet system used in the Bonn polarized proton source[14] shown in fig. 12 illustrates this achromatic focussing of the atomic beam into the ionization volume. A further improvement can be obtained by several separate sextupole or quadrupole magnets, using separate power supplies.

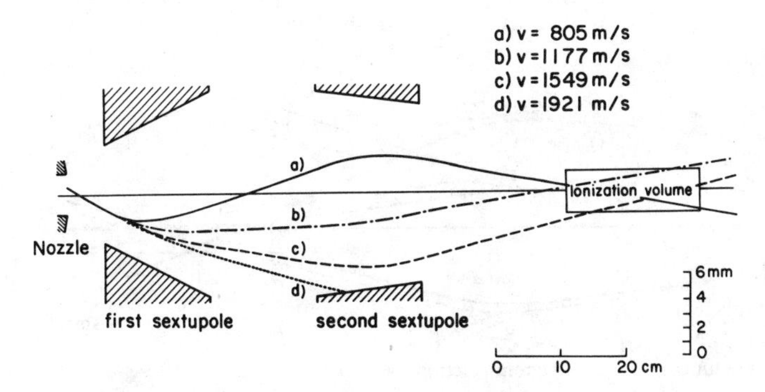

Fig. 12. Achromatic focussing: Calculated trajectories of different velocity in a sextupole doublet system.

The acceptance of the separation magnets and hence the density can be further improved by matching the magnets with the velocity of the atomic beam. A fine demonstration is the investigation by Mathews[14] for the Bonn source. For the different sextupole magnet and rf transition configurations shown in fig. 13 the beam density profile in the ionizer was calculated. The results are shown in fig. 14. It is interesting to notice

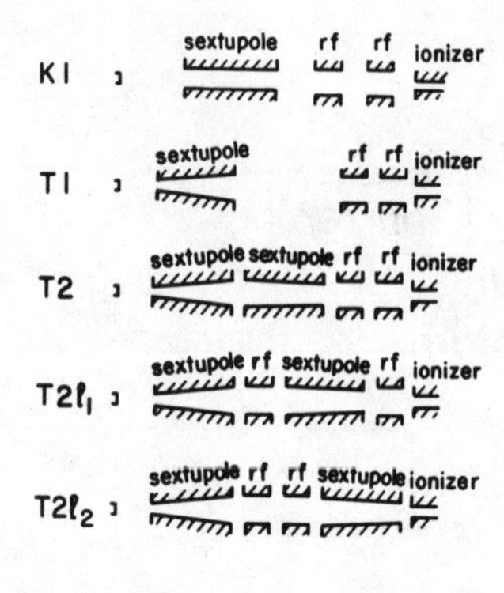

Fig. 13. Sextupole magnet and rf transition configuration.

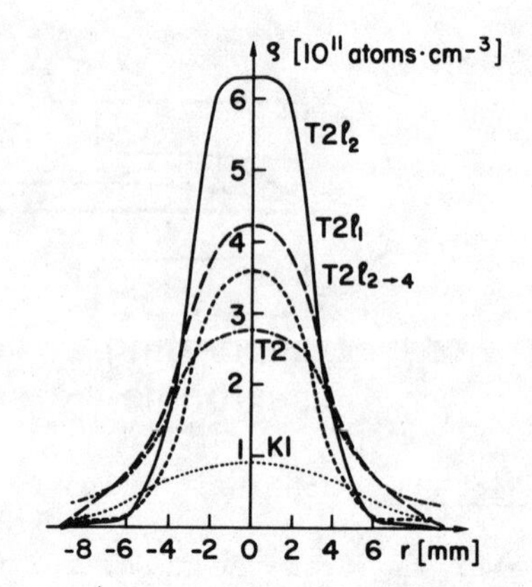

Fig. 14. Calculated density profiles for the configurations of fig. 13.

that the enhancement of the beam density by a more sophisticated magnet configuration can be substantial.

The separate sextupole magnet design allows the introduction of rf transitions between the magnets which results, when activated, in a pure one state atomic beam. While for the 2 to 4 transition a pure state 1 beam enters into the ionizer, a pure state 2 beam can be obtained by a 1 to 3 transition. The resulting trajectories are shown in fig. 15. The loss of a factor two in intensity can be compensated for by the free choice of the optimum magnetic field in the ionizer since any homogeneous magnetic field different from zero can be used.

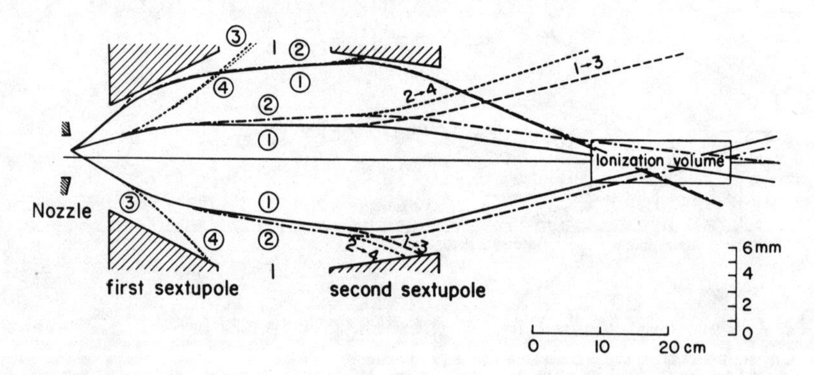

Fig. 15. Calculated trajectories with rf transition between the sextupole magnets for different substates.

Experimentally similar density profiles have been observed. For the CERN polarized hydrogen jet target[17], which uses also a symmetric sextupole pair, the result is shown in fig. 16 for the 2 to 4 rf transition on and off.

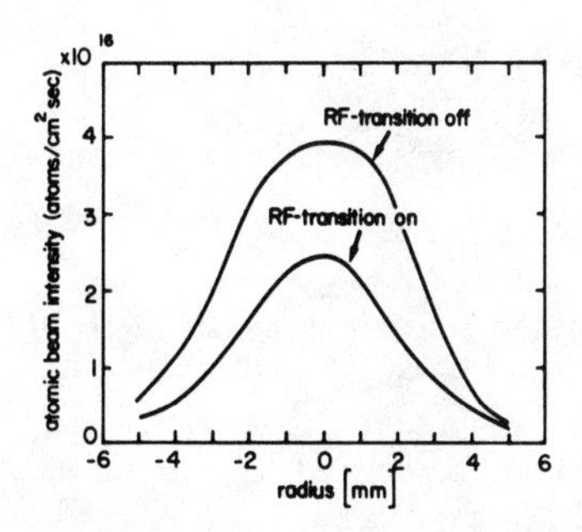

Fig. 16. Measured density profiles with rf transition 2 to 4 between the separation magnet on and off.

The Bonn group has investigated experimentally the density and the density increase for the magnet configurations shown in fig.13. The observed results are presented in fig.17 by the cross hatched bars. The empty bars are derived values for other nozzles or nozzle conditions. The values in the circles on top of the bars are density gains with respect to the magnet configuration K1. The average density is measured on a 3 mm diameter in the ionizer.

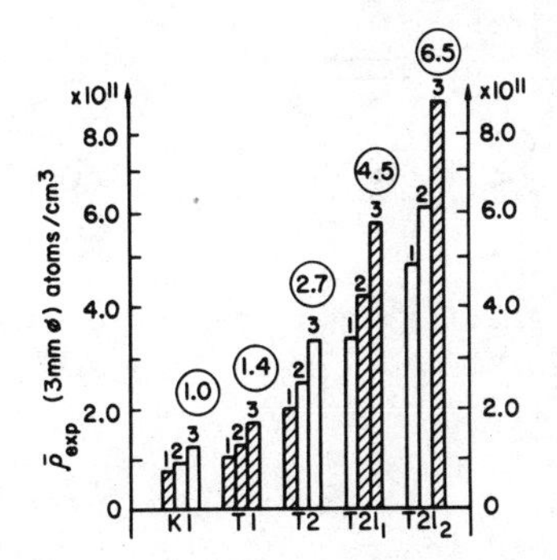

Fig. 17. Density measurements for the magnet configurations shown in fig. 13 on a 3 mm diameter in the ionizer. The cross hatched bars are the result of a measurement. The empty bars are derived values.
1: Laval nozzle SiO_2, 2: Laval nozzle: H_3PO_4 coated, 3: Cylindrical nozzle: H_3PO_4 coated. The values on top of the bars are density gains in respect to the configuration K1.

COOLING OF THE ATOMIC BEAM

The mean velocity of an atom in the beam is

$$v = \sqrt{\frac{2kT}{m}} \tag{7}$$

and the accepted solid angle by a sextupole magnet

$$\Omega = 2.1\,\frac{\mu_B \cdot B_o}{kT} \tag{8}$$

Therefore the density ρ of an atomic beam has a dependence on the temperature like

$$\rho \propto \frac{1}{T^{3/2}} \tag{9}$$

For liquid nitrogen temperature a density gain of 7.5, for liquid hydrogen a gain factor of 57 and for L^4He a gain of 590 results compared with room temperature. These density gains can, however, only be realized, if there is no intensity loss from the dissociator due to the cooling and if the geometry of the beam forming apertures is adjusted adequately. Technically several attempts have been made to reduce the temperature of the beam by cooling the nozzle of the dissociator. Two recent designs are presented in fig. 18. On the left side the Bonn design is shown, which uses a copper nozzle cooled by liquid nitrogen[14]. The recombination in the cooled Cu canal was studied for several treatments. Best results were

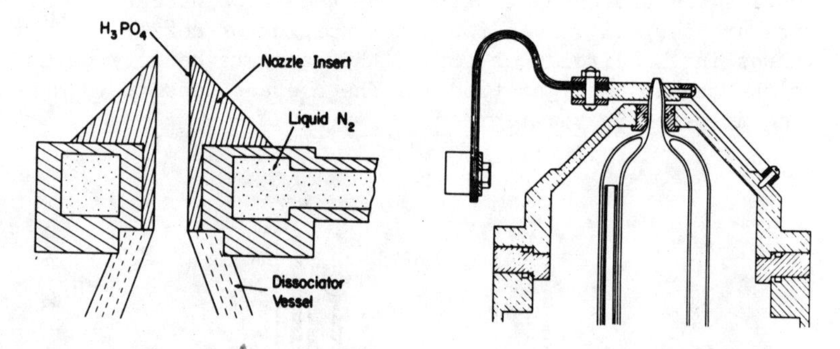

Fig. 18. Nozzle cooling. Left: LN_2 cooled nozzle used at Bonn (ref.[14]) for DC operation. Right: Nozzle cooling (30°K) at ZGS for pulsed operation of the discharge (ref.[18]).

obtained with a H_3PO_4 coating. At Argonne the pyrex nozzle was cooled for pulsed operation of the discharge by contact with a copper block cooled with a closed cycle ^4He refrigerator[18]. The design is shown on the right side of fig. 18. The beam intensity increased by a factor 2.5 when the copper block was held at 28°K. In a earlier attempt at CERN to cool a microwave dissociator the velocity distribution at room temperature and at 77°K were measured. These results are shown in fig. 19 and compared with the Maxwell distribution. The most probable velocity of the beam was found to be about 1500 m s^{-1}, about the same that observed for the Bonn atomic beam. The spread of the velocity distribution is clearly smaller that the expected Maxwellian and indicates that the beam is slightly supersonic. This smaller width is desirable since it reduces chromatic aberration in the sextupole magnet focussing.

An investigation with a cooled rf dissociator was carried out at the ETH some time ago[19]. The technical design is shown by the photograph in fig. 20. The hydrogen gas is fed into the inner pyrex vessel which is surrounded by L N_2. The cooled glass container is insulated by another

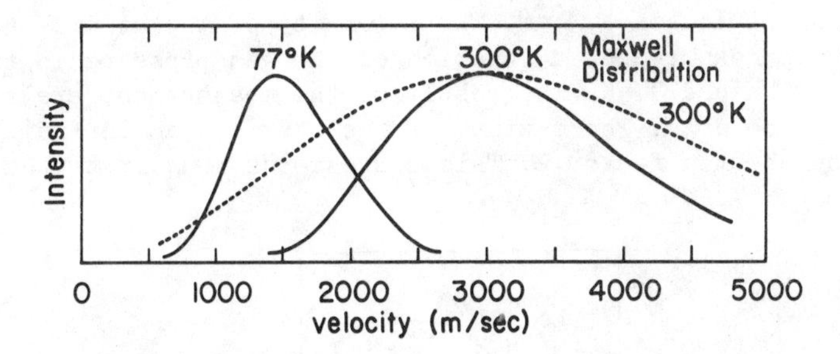

Fig. 19. Velocity distribution of the atomic beam for two different temperatures.

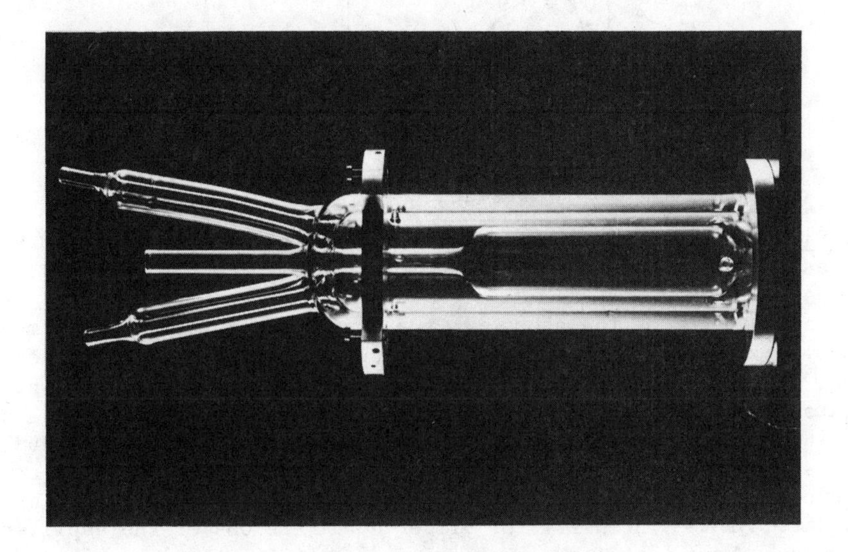

Fig. 20. L N_2 cooled discharge tube.

glass vessel, which is connected to the vacuum container of the atomic beam apparatus. The rf is coupled to the discharge by a coil surrounding the outer glass tube.

Since only a limited pumping capacity was available a low dissociator pressure with a capillary type nozzle was used. The inner diameter of the single capillary was 0,2 mm, the total diameter of the bundle 10 mm, and the length 4.3 mm. The intensity of the atomic beam thus produced was measured as a function of the gas pressure in the discharge tube for 292°K and 77°K. The results of the measurement are shown in fig. 21 by the dots. The maximum of the atomic beam intensity decreases at 77°K by about a factor 2. This corresponds roughly to the decrease of

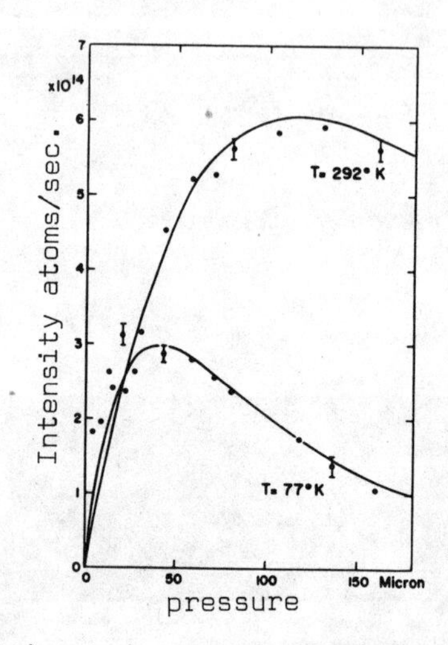

Fig. 21. Measured intensity as a function of the discharge pressure for 77°K and 292°K (dots) and the corresponding calculations (curves).

the mean velocity of the atoms. At the lower temperature the maximum intensity, however, is obtained at about one third of the pressure of 292°K, corresponding approximately to a constant gas density in the dissociator tube. This observed behaviour was calculated by a model, which took into account the scattering in the capillaries and the residual gas. The results of the calculation are shown in fig. 21 by the solid curves. The success of this result justifies the use of this method for the estimation of the intensity for different dimensions of the capillaries. The result shows that for a LN_2 cooled dissociator the intensity of the atomic beam can be even increased provided the geometry of the capillaries are chosen adequately[19]. Hence the final gain of the ion beam depends only on the acceptance of the separation magnet and the velocity of the beam. This is a very encouraging result since in this case relation (9) applies.

The moderate success in increasing the beam intensity by cooling the beam raises the question for its causes. One has to remember that all the investigations have been made with a fixed beam-forming geometry and magnets designed for a 300°K beam. As can be seen in the example of the ETH atomic beam geometry in fig. 22 the opening angle is ±2.5° and this is also the angle for which the magnet is designed. Cooling the beam alone does not change substantially the acceptance angle as easily can

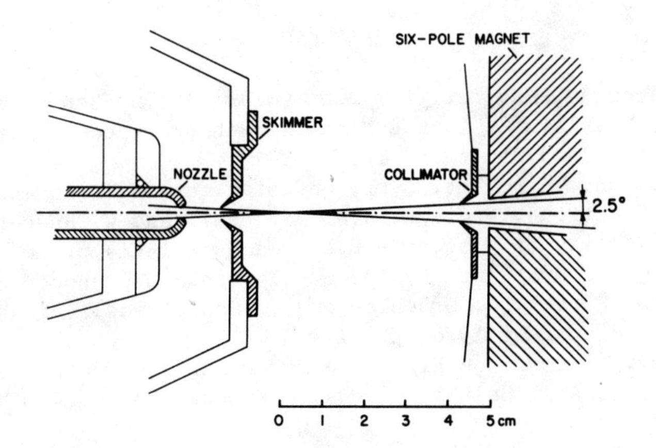

Fig. 22. Geometry of the beam forming apertures.

be observed from the trajectories of the atoms in the figs. 5, 8 and 9 for an entrance angle of 3°. The diameter of the beam envelope decreases with decreasing temperature but for a fixed geometry the available space in the magnet is no longer optimally used. In order to obtain the full gain a complete new design of the beam-forming and the magnet poles is required. In this way the complete benefit of the cooling can be expected.

CONCLUSIONS

While in the last few years the development of the ionizing stage of atomic ground state polarized ion sources has made rapid progress, the atomic beam stage has improved only slowly. The production of highly-polarized H^- beams which are most promising for the acceleration in high energy accelerators is most advanced in two laboratories, which even are using two different ionization schemes[8,10], but both are based on the ionization of thermal atomic hydrogen beams. In the future an increase of the present 100 μA polarized H^+ and 3 μA H^- beams by nearly two orders of magnitude seems to be possible by the cooling of the atomic beam to low temperature and the design of a suitable multistage separation magnet system. This development requires, however, a systematic investigation of the cooling and the formation parameters of the atomic beam, as carried out for room temperature by Risler et al.[20], and the systematic examination of possible separation magnet designs. Since not only dissociator and magnet system are interrelated but also the ionizer and charge exchange device, the characteristics of all these components has to be taken into account for a final design of a high intensity polarized proton source. For example a new design of the sodium charge exchange canal of the ETH source lead to a better matching of the available positive beam and hence to the expected gain of a factor two in the negative beam intensity as predicted in ref.[10].

The present review shows the possible directions of future developments, the schedule of the progress is however more difficult to estimate.

REFERENCES

1. R. Fleischmann, 2nd Int. Symp. on Polarization Phenomena of Nucleons, Karlsruhe, P. Huber and H. Schopper, eds. (Birkhäuser Verlag Basel and Stuttgart 1966) p.21
2. H.F. Glavish, Third Int. Symp. on Polarization Phenomena in Nuclear Reactions, Madison, H.H. Barschall and W. Haeberli eds. (The University of Wisconsin Press Madison 1971) p. 267
3. T.B. Clegg, Fourth Int. Symp. on Polarization Phenomena in Nuclear Reactions, Zürich, W. Grüebler and V. König, eds. (Birkhäuser Verlag Basel and Stuttgart 1976) p. 111
4. H.F. Glavish, Conf. on Higher Energy Polarized Proton Beams, A.D. Krisch, A.J. Salthouse, eds. AIP Conf. No 42 (AIP New York 1978) p. 136
5. W. Grüebler, Proc. Fifth Int. Symp. on Polarization Phenomena in Nuclear Physics, Santa Fe, G.G. Ohlsen et al., eds. AIP Proc. No 69 (AIP New York 1981) p. 848
6. W. Haeberli, Proc. Int. Symp. High Energy Physics with Polarized Beams and Polarized Targets, Lausanne, C. Joseph and J. Soffer, eds. (Birkhäuser Basel 1981) p. 199
7. B.L. Donnally and W. Sawyer, Phys. Rev. Lett. 15, 439 (1965)
8. D. Hennies et al., Phys. Rev. Lett. 40, 1234 (1978)
9. H.F. Glavish, IEEE Trans. on Nucl. Science 26 1517 (1979)
10. P.A. Schmelzbach, W. Grüebler, V. König and B. Jenny, Nucl. Instr. Meth. 186, 655 (1981)
11. W. Grüebler, P.A. Schmelzbach, V. König and P. Marmier, Phys. Lett. 29A 440 (1969)
12. W. Grüebler, P.A. Schmelzbach, V. König and P. Marmier, Helv. Phys. Acta 43 254 (1970)
13. A.S. Schlachter, K.R. Stalder and J.W. Sterns, Phys. Rev. A22 2494 (1980)
14. H.G. Mathews, Ph.D. thesis, University of Bonn (1979)
15. G.J. Witteveen, Nucl. Instr. Meth. 158 57 (1979)
16. H.F. Glavish, Proc. Fourth Int. Symp. on Polarization Phenomena in Nuclear Reactions, W. Grüebler and V. König, eds. (Birkhäuser Verlag Basel and Stuttgart 1976) p. 844
17. L. Dick, J.B. Jeanneret, W. Kubischta and J. Antille, Proc. Int. Symp. High Energy Physics with Polarized Beam and Polarized Targets, Lausanne. C. Joseph and J. Soffer eds. (Birhäuser Verlag Basel and Stuttgart 1981) p. 212
18. P.F. Schultz, E.F. Parker and J.J. Madsen, Proc. Fifth Int. Symp. on Polarization Phenomena in Nuclear Physics, Santa Fe, G.G. Ohlsen et al. eds. AIP Proc. No. 69 (AIP New York 1981) p. 909
19. M. Perrenoud, W. Grüebler and V. König, Helv. Phys. Acta 44 594 (1971)
20. R. Risler, W. Grüebler, V. König and P.A. Schmelzbach, Nucl. Inst. Meth. 121, 425 (1974)

PROGRESS REPORT ON IMPROVED ATOMIC BEAM STAGE
(Working Group Summary)

W. Grüebler
Laboratorium für Kernphysik, Eidgenössische Technische Hochschule,
8093 Zürich, Switzerland

INTRODUCTION

This is partly a summary of the working group opinion of the status and the potential for improving the ground state atomic beam stage of polarized ion sources. Partly the content also contains my own consideration stimulated by the workshop and is based on additional work carried out after the workshop.

The requirements for a modern atomic hydrogen beam have been discussed in the review paper of the atomic beam stage of ground state polarized proton ion sources of the present workshop[1]. In a conventional atomic beam apparatus electron polarization and consequent nuclear polarization by rf transition is easily obtained with polarization values of more than 95%. A convenient set of rf transitions allows the change of the sign of the nuclear polarization with only very small intensity modulation ($<10^{-4}$) and without substantial change of the optical properties of the polarized ion beam. The remaining problem concerns the intensity of the ion beam, which is mainly controlled by the intensity and density of the atomic beam reaching the ionizer and the matching of the atomic beam with the corresponding ionizing electron or neutral Cs beam. Part of the matching problems are due to the large velocity and the broad velocity distribution of the atoms in the atomic beam. Prominent problems in producing a high intensity polarized atomic beam are the dissociation of the molecular hydrogen and the formation of an atomic beam.

The following topics were discussed in several sessions by the working group for an improved atomic beam stage.

1) Gas dynamics and nozzle design and their influence on beam formation.
2) Atomic beam parameters and their limiting factors.
3) Dissociator cooling, reduction of beam velocity spread and dissociation techniques.
4) Sextupole configurations including magnetic field gradients.

BEAM FORMATION BY A NOZZLE

For atomic hydrogen gas with a density $n = 6 \cdot 10^{14}$ atoms cm^{-3}, which is equivalent to a pressure of $2 \cdot 10^{-2}$ torr at room temperature, the mean free collision path λ is approximatly 3 mm. This gives a Knudson number $Kn = \frac{\lambda}{d}$ of about 1 for a nozzle diameter d of a few mm. Hence we have effusion in the molecular flow regime. The maximum rate Q of atoms that emerge from a thin-walled aperture is

$$Q = \frac{1}{4} \cdot n \cdot \bar{v} \cdot A \qquad (1)$$

where $\bar{v}$ is the mean atomic velocity inside the dissociator tube and A is the area of the aperture. Under such conditions the effusion has a cosine distribution. If the thin-walled aperture is replaced by a canal-like nozzle the angular distribution of the emergent beam is changed considerably. The intensity is more concentrated around the center line and the directionality is increased.

Taking the above density and the mean velocity at 300°K, $\bar{v} = 2.5 \cdot 10^5$ cm sec^{-1}, one obtains a maximum intensity per mm^2 area of

$$Q = \frac{1}{4} \cdot 6 \cdot 10^{14} \cdot 2.5 \cdot 10^5 \cdot 10^{-2} = 3.75 \cdot 10^{17} \text{ atoms sec}^{-1}.$$

The acceptance angle of a present sextupole magnet is about ± 2.3° i.e. at most a fraction of about 10^{-2} of the cosine distribution is accepted by the separation magnet. Thus a maximum beam intensity of 10^{15} atoms sec^{-1} can reach the ionizer.

Modern atomic beam devices however work with a dissociator pressure of several torr. This corresponds to a density of 10^{17} atoms cm^{-3}. In this case $Kn < 0.01$ and the molecular flow description no longer applies. In fact one has viscous flow or one is in the transition region between molecular and viscous flow. Since a gas dynamical beam is formed, the relation of this type of regime should be used to calculate the throughput. Assuming a inviscid flow, which is approximatly justified for a simple aperture or a short canal, the particle current leaving the dissociator is

$$Q = p_o \cdot \frac{2\gamma}{R\,T_o(\gamma+1)} \cdot \left(\frac{2}{\gamma+1}\right)^{\frac{1}{\gamma-1}} \cdot A_{min} \qquad (2)$$

$$= n_o \cdot \frac{2 \cdot R \cdot T_o \cdot \gamma}{\gamma + 1} \cdot \left(\frac{2}{\gamma+1}\right)^{\frac{1}{\gamma-1}} \cdot A_{min} \qquad (2a)$$

where p_o, n_o, T_o are the pressure, density and temperature in the discharge tube, γ is the specific heat ratio and A_{min} is the minimum area of the aperture. Here the throughput reaches a constant value if the pressure outside the nozzle is below the critical pressure p_{cr}. For atomic gases $\gamma = 1.66$ and the ratio $p_{cr}/p_o \simeq 0.5$. If this condition is obtained the atoms in the smallest area A_{min} reach the velocity of sound. A supersonic flow can be achieved if a converging-diverging Laval nozzle is used. Because of the low density in the present dissociator the boundary layer in the diverging part is very thick and hence will encroach on the isentropic core and reduce the area A_{min}. In fact the investigation of this problem shows that no benefit from a diverging part of a Laval nozzle can be found. A truly profitable supersonic flow occurs at a pressure $p_o > 10$ torr at room temperature, in which case experiment agrees with the supersonic flow description. Unfortunatly at this high density the volume recombination of atomic hydrogen is already considerable and overrules the benefit of the supersonic flow.

DEPENDENCE OF THE ATOMIC BEAM ON THE TEMPERATURE

Hot atomic beam

Although it is generally accepted that a cooler atomic beam (with atoms at lower velocity) can yield larger polarized ion beams, it was pointed out that intense H^o sources at 10 eV were now available. Therefore the question was raised, whether the ion intensity would benefit by such a fast atomic beam being separated in a sextupole magnet and subsequently ionized by surface ionization to H^-ions. The process $\vec{H}^o$ + surface $\rightarrow \vec{H}^-$ has an efficiency of about 5%. A simple calculation shows that the increase from thermal energy ($T = 300^o \approx 0.03$ eV) to 10 eV decreases the accepted solid angle from typically $3 \cdot 10^{-3}$ sr ($\pm$ 2.3°) to about 10^{-5} sr (half angle ∿0.1°) which is a factor 400. Therefore for the same differential intensity flux from the source this factor 400 would have to be gained by a higher ionization efficiency. For a break-even point a 40% efficiency would be required, which is about a factor 10 beyond present capabilities.

Dissociator cooling

The advantages of cooling the dissociator at the nozzle in order to improve the atomic beam density in the ionizer has been discussed extensively in the review talk of the present workshop. The problems involved in such cooling are also discussed in the working group session. A typical example is the Argonne pyrex nozzle cooling using a closed-cycle He refrigerator, resulting in a measured temperature of 30° at the copper block surrounding the nozzle. This cooling led to reduce rf power requirements for the dissociator, lower current in the sextupole magnet coils and increase of atomic beam intensity by a factor 2.5. The results of the velocity measurements made on the atomic beam stage are given in the following table.

Measured temp. °K	Most probable velocity of the atoms $\bar{v}$[cm sec^{-1}]	Velocity spread Δv[cm sec^{-1}]
300	$2.8 \cdot 10^5$	$1.2 \cdot 10^5$
30	$1.7 \cdot 10^5$	$0.5 \cdot 10^5$

The velocity $1.7 \cdot 10^5$ cm sec^{-1} corresponds to a temperature of 110°K. The increasedintensity suggests that already a simple cooling of the nozzle can bring a substantial benefit. The reduced magnet current however shows that the beam forming geometry and the separation magnet has to be adapted to the new mean velocity. Apparently this has not be done for the ZGS source and therefore only part of the possible profit could be obtained. On the other hand the measured most probable velocity of the atoms shows that the cooling was not fully adequate, which results in a loss of obtainable intensity. The critical problem of the cooling of the atoms in the nozzle is further illuminated by the observation of the Bonn group for two different lengths of the nozzle (6mm and 14mm)[2]. This is shown in fig. 1, where the atomic beam intensity is plotted versus the current in the second sextupole magnet (compressor magnet). The current of the

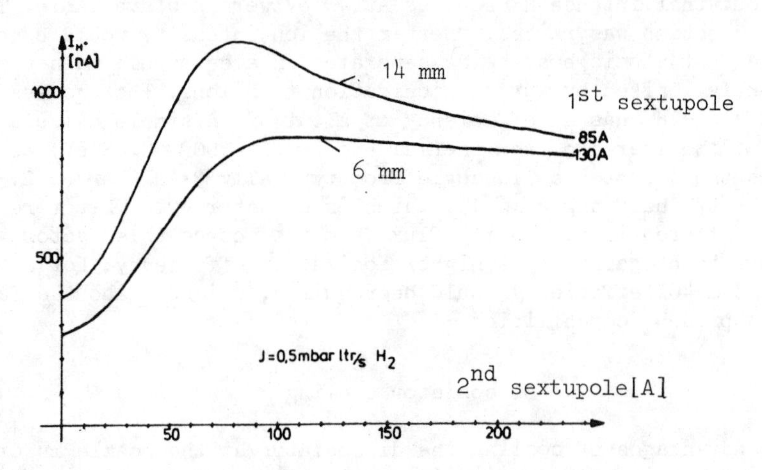

Fig. 1. Atomic beam intensity versus the current in the compressor sextupole magnet for two lengths of the nozzle.

first sextupole magnet is held constant at 85 A and 130 A respectively, the optimum values for maximum beam intensity. Clearly the longer nozzle canal is substantially superior to the shorter one, which is attributed to the better cooling of the atomic beam. For this reason it was suggested that a H_3PO_4 coated copper nozzle as at Bonn or a sapphire nozzle as used at CERN could provide a better cooling of the beam. Finally the question arises if the method shown in fig. 2 will not be the final answer to the cooling problem. Here the molecular hydrogen

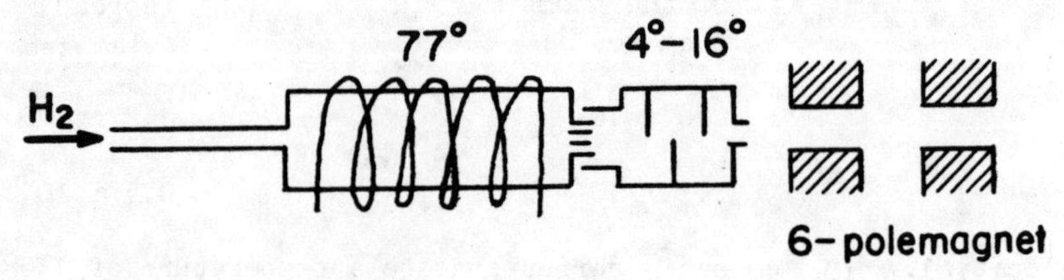

Fig. 2. Proposal for a cooled dissociator.

gas is dissociated in a vessel at 77° by an rf discharge. The technical feasibility of this scheme was proved already a decade ago[3]. Subsequent cooling to lower temperature, preferably below the freezing point of molecular hydrogen in order to prevent surface recombination could be accomplished in a simple container with a beam forming nozzle.

For a constant pressure p_o in the nozzle container the total intensity of the atomic beam changes like $1/\sqrt{T}$ as shown by the relation (2). In this case the atomic gas density increases proportionally to T. This density may however be limited by volume recombination. If one has already reached in present dissociators the maximum allowed density and the volume recombination rate does not decreases with decreasing temperature, then one has to keep the density constant and relation (2a) applies. Thus the atomic beam intensity drops proportionally to $\sqrt{T}$ for cooler beams.

SEPARATION MAGNET

Multipole magnets are used for the separation of the substates 1 and 2 from 3 and 4. Preference in most present sources is given to sextupole magnets because of the optical properties. The focal length is given by the gradient of the magnetic field and the velocity of the atoms. For a given radius of the pole tip from the axis, the gradient is limited by the B field on the pole tip because of the saturation of the magnetic material. At the present time electromagnets are used. Conventional sextupole magnets yield 0.7 - 0.9 T on the pole tips. A value 1.2 T seems to be an upper limit for $B_{pole\ tip}$. The use of a new type of permanent multipole magnet made from a Sm-Co alloy[4] should be tried in future sources, in particular for beams with low velocity (cooled beams). A value of $B_{pole\ tip} = 1.3$ T is reported from a prototype magnet at CERN. Fig. 3 shows a schematic of this type of magnet. Another type of sextupole magnet could be made using superconducting coils. A computer model by K. Kubischta yields 2.5 T on the pole tips with an aperture of 2 cm diameter.

The reduction of the mean velocity of the atoms by cooling the dissociator or the nozzle requires a complete redesign of the conventional magnet in order to take full advantage of the cooling. With present

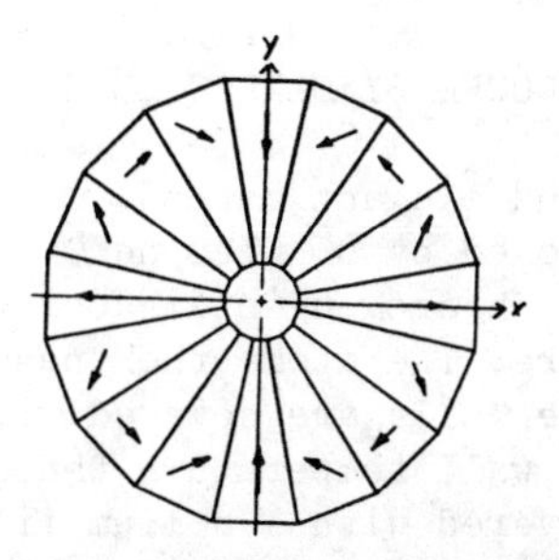

Fig. 3. Schematic cross section of a 16-piece REC quadrupole.

techniques a half angle of about 10°, corresponding to a solid angle of 10^{-1}sr, can be obtained. This is a gain of a factor 20 compared to the solid angle of present sources. A realistic design to scale is shown in fig. 4 for atoms corresponding to a temperature of 4^oK(upper part) and 10^oK(lower part). The distance between nozzle and skimmer is held practically constant: the distance between skimmer and collimator is reduced by about 30%. It is well known that this distance is not very critical for

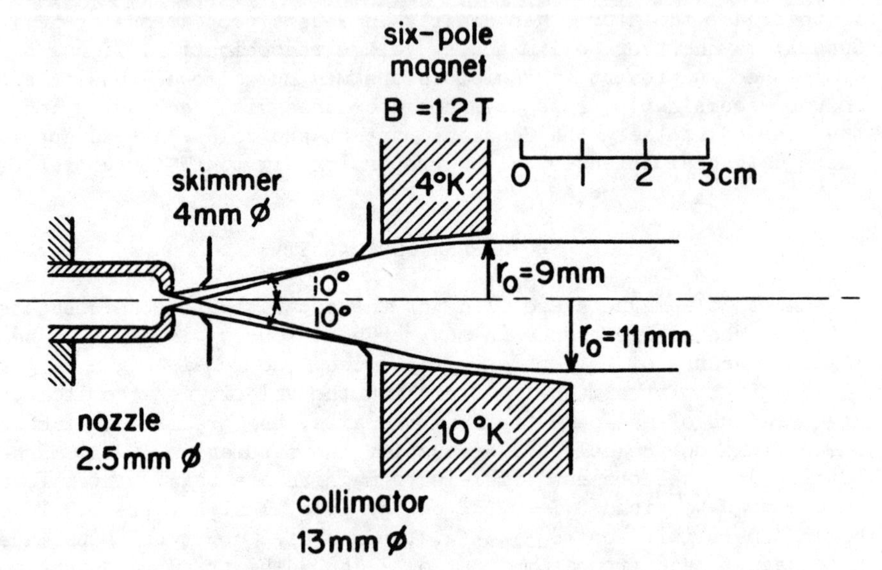

Fig. 4. Design for a atomic beam apparatus (to seale) for cooled atoms. Upper part for 4^oK, lower part for 10^oK.

the beam formation[5]. Thus all the conductances are basically not changed The lengths of the magnets in fig. 4 are chosen to obtain a parallel atomic beam. For the focussing in the ionizing region a second magnet is required. The interesting feature of such separation magnets is their shortness. The short length can also be provided by the new type of permanent magnet shown in fig. 3.

ATOMIC BEAM PRODUCED BY ULTRACOLD HYDROGEN ATOMS

A new, interesting and elegant method to produce a high intensity atomic beam has been proposed by D. Kleppner[6]. A schematic diagram is shown in fig. 5. Molecular hydrogen is dissociated by an rf discharge at liquid nitrogen temperature. The atoms then pass through a baffled apertu into a 4.2^oK chamber whose walls are covered with frozen molecular hydrogen. After cooling to the wall temperature the atoms enter a cell cooled to 0.5^oK. This cell is covered with a helium film in order to prevent sur face recombination. The complete set-up is contained in a strong field

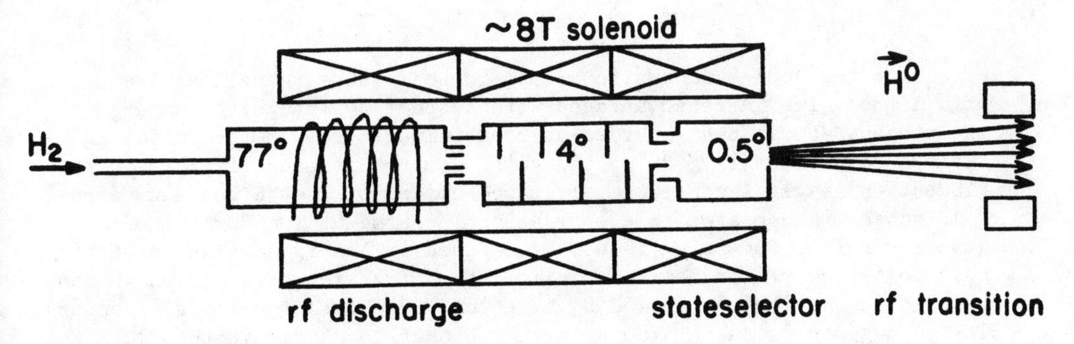

Fig. 5. Proposal for an ultracold hydrogen atomic beam.

of 8 T produced by a superconducting solenoid. The electron spin polarization is achieved at the end of the magnetic field where a nozzle of the coldest cell is located. In this inhomogeneous fringe field the states 3 and 4 are repelled back into the strong magnetic field, forming a magnetic bottle. The states 1 and 2 are ejected and accelerated due to the magnetic potential. These spin selected atoms have a velocity corresponding to about 6 - 8°K. The gain of velocity is mainly in the forward direction, therefore the intensity is concentrated around the center line. This is graphically shown in fig. 6. Since the transverse component stays constant, the beam has much more directionality than the cosine distribution of an effusion source. For this reason also the fractional

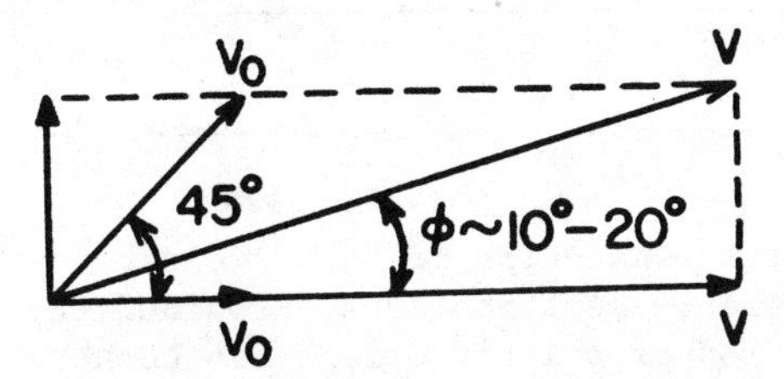

Fig. 6. Directionality of the ultracold beam.

velocity spread is quite small. In fact the velocity is nearly monochromatic: the spread Δv at 7°K is the same as for the Maxwellian distribution at 0.5°K. This feature is shown in fig. 7. Nuclear polarization finally is obtained by a set of rf transitions before the entrance to an ionizer.

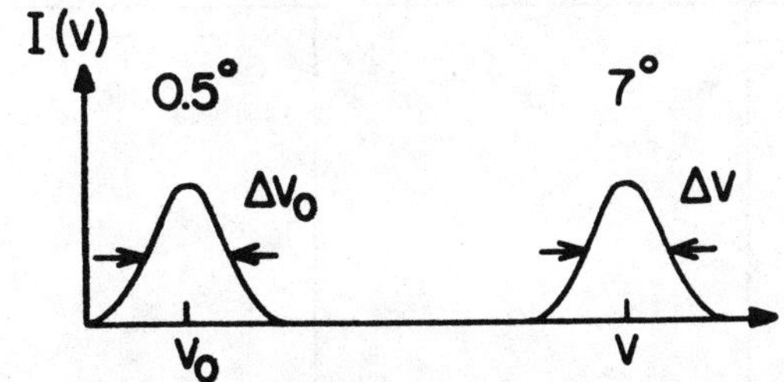

Fig. 7. Velocity spread of the ultracold beam.

COMPARISON AND CONCLUSIONS

Both the cooled conventional atomic beam stage as well as the ultracold method give large advantages with respect to intensity compared to the present atomic beam sources at room temperature. In the following table a comparison is made of the gains due to different factors in the production stage. A conventional source cooled to $T = 4^{o}K$ and the ultracold scheme is compared to a conventional source at $T = 300^{o}K$. The quantity f is a focussing factor as defined in the first session of the cold polarized proton source working group (fraction of atoms which can be used in the ionizer including a factor 1/2 due to the state selector). The throughput is calculated by assuming that no change in the flow regime occurs and the density in the nozzle cell is constant.

	Conventional source $T=300^{o}K$	Conventional source $T=4^{o}K$	Gain	Ultracold source	Gain
throughput Q	1	1/9	1/9	1/25	1/25
solid angle accepted Ω sr	$5 \cdot 10^{-3}$	10^{-1}	20	0.5	100
focussing factor f	0.3	0.3	1	1	3
relative ionizer efficiency	1	9	9	7	7
total gain of H^{+}	1		20		84

It is likely that the matching of these cold beams in the ionizer requires a redesign of the present ionizers. Provided the present high efficiency for the production of polarized ions can be kept, the following polarized hydrogen ion beams can be obtained.

	Conventional source $T=300^{o}K$ [7,8]	Conventional source $T=4^{o}K$	Ultracold source
H^{+} (μA)	100	2000	8400
H^{-} (μA)	3	60	250
improved $H^{+} \rightarrow H^{-}$ charge exchange[9] H^{-} (μA)	20	400	1700

This summary shows clearly that for ground state atomic beam sources there is still a large potential for improvements, which has barely started to be investigated. In particular the atomic beam stage shows excellent promises for a large increase of atomic beam intensity and density either by the conventional atomic beam technique or by the new scheme of ultracold beams. While the above tables exhibit amazingly high values of gains and absolute values of positive and negative ion intensities one should be aware that these are projected values. Not all the technical problems are yet solved and as we can learn from the past, the solutions sometimes come quite slowly and require tenacious work. With this experience in mind we should be realistic for the future and use also a time projection for the next step in the development.

REFERENCES

1) W. Grüebler, Review of Atomic Beam Stage of Ground State PPIS, this workshop p. 51 .
2) H.G. Mathews, PhD thesis, University of Bonn (1979)
3) M. Perrenoud, W. Grüebler, and V. König, Helv. Phys. Acta 44 (1971) 594
4) K. Halbach, Nucl. Instr. and Meth. 169 (1980) 1
5) R. Risler, W. Grüebler, V. König and P.A. Schmelzbach, Nucl. Instr. and Meth. 121 (1974) 425
6) D. Kleppner, Review of Possible Uses of cold Atomic Hydrogen in PPIS, this workshop p. 105 .
7) P.A. Schmelzbach, W. Grüebler, V. König und B. Jenny, Nucl. Instr. and Meth. ___ (1981)
8) D. Hennies et al., Phys. Rev. Lett. 40 (1978) 1234
9) see report on $H^+ \to H^-$ Ionization Working Group, this workshop p. 103 .

Report on the Improved Atomic Beam Stage Work Session I

Wednesday, May 20, 1981

Chairman: W. Gruebler   Secretary: S.R. Magill

The topics suggested for discussion were the following:
1) Atomic beam parameters and their limiting factors;
2) Dissociator cooling, reduction of beam velocity spread, and dissociation techniques including RF, microwaves, and lasers;
3) Sextupole configurations including magnetic field gradients and the use of superconducting coils; and
4) Effect of collisions in beam formation and their influence on gas dynamics, nozzle design, and vacuum pumping speed.

Summaries of the discussion of each topic and any conclusions reached by the working group follow.

1. Atomic Beam Parameters - The two parameters discussed were the energy range of the atomic beam and the desired atomic density. With regard to density, it was mentioned that 10^{12}-10^{13} cm^{-3} is desired, but the currently attained value is ~ 7×10^{11} cm^{-3}. It was generally accepted that lower velocity atomic beams formed by cooling the dissociator nozzle were desirable. Then the question was raised: Can one use a much faster (e.g. 10eV) H° beam with a sextupole magnet in order to increase ionization efficiency? It was pointed out that intense sources at 10 eV were now available. W. Haeberli worked out the following calculation as an answer to this question:

The solid angle accepted by the sextupole magnet is given by

$$\Delta\Omega \sim \mu B_0/kT$$

or typically 3×10^{-3} sr. This corresponds to a half angle of ~2°. If 2×10^{16} atoms sec^{-1} pass through the sextupole, the differential intensity is $2\times10^{16}/3\times10^{-3} \cong 7\times10^{18}$ atoms sr^{-1} sec^{-1}. If the nozzle aperture is ~ 3 mm^2 (typical), the differential intensity flux is ~2×10^{18} atoms mm^{-2} sr^{-1} sec^{-1}. If the beam energy is raised from 300°K (~ .025 eV) to 10eV, the acceptance solid angle of the sextupole is smaller by a factor of ~ 400 (corresponding to a half angle of 0.1°). Therefore, for the same differential intensity flux from the source, a factor of 400 would have to be gained in ionization efficiency which for H^- is $\varepsilon \sim 10^{-3}$ for present sources. This means that the ionization efficiency for H^- would have to be 50%, yielding a beam flux of ~ 0.3 amps mm^{-2} sr^{-1} sec^{-1}. This is a factor of ~10 beyond current capabilities.

2. Dissociator cooling, velocity spread reductions, and dissociation techniques -

0094-243X/82/80079-05 $3.00 Copyright 1982 American Institute of Physics

a. Argonne dissociator cooling -Cooling of the dissociator nozzle of the ZGS source led to a factor of ~ 2.5 increase in atomic beam intensity, reduced RF power requirements, and a lower sextupole current. Velocity measurements made on the atomic beam stage cooled by a closed cycle liquid He refrigerator are shown in the following table:

Atomic Beam	Temp. (°K)	Beam Velocity (most probable)	Δv
H	300	2.8 km/sec	1.2 km/sec
H	30*	1.7 km/sec	0.5 km/sec
H_2	30*	0.8 km/sec (RF on)	
H_2	30*	1.5 km/sec (RF off)	

* temperature measured at the copper block surrounding the pyrex dissociator nozzle.

The temperature of the beam itself was not measured, but the value of 1.7 km/sec corresponds to a temperature of somewhat greater than 100°K. It was suggested that the H_3PO_4 nozzle used at Bonn or a sapphire coated nozzle as used at CERN could provide a better method of coupling cooling power to the beam, resulting in lower temperatures. It was also noted that for the pulsed ZGS source, dissociation seems to occur at the nozzle tip.

b. CERN dissociator - The following graphs show experience with pulsed RF dissociator for various beams:

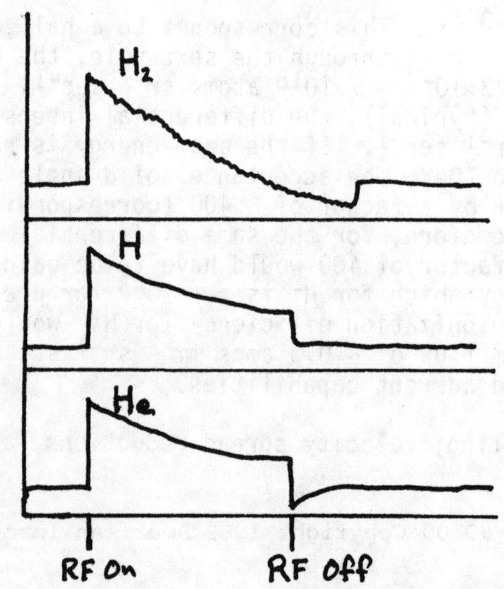

An experiment was also done by varying the length of the capacitively coupled RF cavity around the position corresponding to $\lambda/4$. This indicated that dissociation of H_2 was occuring within the dissociator bottle.

c. SIN-Zurich dissociator -Dissociator bottle parameters given were ~ 10 mm O.D. and ~ 20 cm length. The RF parameters were 250 Watts of capacitively coupled RF at 27 MHz. When cooling the dissociator, the beam intensity seemed to saturate at ~ 100°K. Further cooling provided little increase in beam intensity.

Other ideas discussed under this topic were the following:

-What about separating the dissociation and cooling processes? A method was suggested to cool the gas and then dissociate with a laser by frequency doubling or tripling. This idea is being investigated and will be reported on in a later working group.

-Another question was asked about the importance of various dissociator nozzle parameters such as tilting off axis, distance from the skimmer, diameter and length of nozzle, and others. The answer to this question was indefinite and led to the following conclusion regarding the cooling and dissociation properties of the atomic beam stage. The working group concluded that a systematic program was needed in order to determine the effect of cooling on beam intensity, optimization of dissociator techniques, and the dependence of both of these on parameters of the dissociator nozzle.

Suggested topics 3) and 4) were not addressed in this work session.

Report on the Improved Atomic Beam Stage Work Session II

Monday, May 25, 1981

Chairman: W. Gruebler Secretary: S.R. Magill

Discussion continued in this working group on the remaining topics listed at the first meeting, namely, 3) Sextupole magnet configurations, and 4) Dissociator nozzle gas dynamics.

3. Sextupole Magnets - It was reported that the B-field upper limit at the pole tip of a normal conducting sextupole (soft iron tips) is ~ 1.2T. The ANAC models yield 0.7→0.9T. W. Kubishta reported on the development of permanent magnet sextupoles at CERN. The following reference was given:

K. Halboch, NIM, 169 (1980) p. 1-10.

These magnets were made from a Sm-Co alloy which has a B_r=0.95T. The following equation can be used to calculate the B-field at the pole tip:

$$B_{pole\ tip} = B_r \left(\frac{3}{2}\right)\left(1 - \left(\frac{r_1}{r_2}\right)^2\right)$$

where r_1 = inner radius (aperture size), and r_2 = outer radius (magnet cross-section). The magnets at CERN were made with r_1/r_2 = 1/4, yielding $B_{pole\ tip}$ = 1.3T (measured with Hall probe). A method was devised for tuning these magnets by constructing two concentric magnets and turning against each other until B=0 at center. It was suggested that the magnets could be placed closer to the nozzle to increase acceptance angle. Also, with higher B-field or cooler beam, the magnet length would decrease, and pumping may then be done after the magnet. It was thought that skimmers and differential pumping would still be needed at the nozzle for a DC source, but probably a pulsed source would not produce the high gas load, and so this idea may be feasible. It was also suggested that permanent magnets might not be attractive because they cannot be changed after production. If, however, properties of the dissociator are well known (velocity, Δv, etc.), designs can be made with confidence.

Another type of sextupole could be made from superconducting coils. A computer model by W. Kubishta produced $B_{pole\ tip}$ ~ 2.5T with an aperture of 2 cm diameter. These configurations could be attractive due to increased flexibility (over permanent magnets) and the high B-field generated.

4. Dissociator Nozzle Design - So far, gas dynamics at the nozzle are not well understood. It was concluded that a systematic study of

these gas flow parameters is needed in order to understand nozzle operation and to design an optimum system. One question to be answered is that of an upper limit for the density of atomic beam attainable in the nozzle. At this time, there is no reason to believe that present densities obtained are limited. One problem at higher densities may be that recombination could dominate in the dissociator, thus reducing the atomic beam. These ideas should be investigated in more detail.

For the short Improved Atomic Beam Stage Work Session III, see the joint session under Cold Atomic Hydrogen (page 135)

IONIZATION OF THERMAL POLARIZED HYDROGEN ATOMS

W. Haeberli
University of Wisconsin, Madison, WI. 53706

ABSTRACT

The ionization stage of atomic-beam polarized ion sources is reviewed. Recent progress in electron-bombardment ionizers has led to D.C. $\vec{H}^+$ beams of 100 μA. Charge exchange in Na vapor at 5 keV energy has permitted production of 3 μA $\vec{H}^-$ ions. Direct conversion of $\vec{H}^o$ to $\vec{H}^-$ by the colliding-beam method using a 40 keV beam of Cs^o has been shown to yield 3 μA of $\vec{H}^-$. The possibility of surface ionization is briefly mentioned. If the best presently known techniques are combined, pulsed $\vec{H}^+$ beam currents of 400 μA and pulsed $\vec{H}^-$ currents of 20 μA are anticipated. Further development work promises additional substantial gains.

1. INTRODUCTION

Atomic-beam polarized ion sources consist of two principal parts: an atomic-beam apparatus that produces a directed beam of thermal polarized hydrogen atoms, and an ionizer in which the $\vec{H}^o$ atoms are converted into $\vec{H}^+$ or $\vec{H}^-$. Although production of $\vec{H}^-$ is less efficient than production of $\vec{H}^+$, beams of negative ions are of interest to high energy accelerators because of the gain that results from multiturn injection.

To simplify the following discussion, we will assume a conventional atomic beam apparatus that produces 2×10^{16} $\vec{H}^o$/sec with an average velocity of $\bar{V} = 3 \times 10^5$ cm/sec. It can be assumed that the atomic beam is sufficiently well collimated (if necessary by use of a compressor six pole) to be confined to a diameter of 1 cm over a length of 10 to 30 cm. This translates into an average volume density $\rho = 10^{11}$ H^o/cm^3 (corresponding to a pressure of 2.5×10^{-6} Torr), and a linear density $dN/dz = 7 \times 10^{11}$/cm, where z is the momentum direction of the atoms. These numbers were chosen because they are typical of atomic beam sources now in use, but it must be emphasized that they do not present the best values attainable with current methods. In particular, ρ can be increased considerably, e.g. by cooling the dissociator nozzle.

Another assumption we will make is that ionization of the $\vec{H}^o$ atoms takes place in a "strong" magnetic field, typically B=0.2T. This requirement arises from the need to decouple electron and nuclear spin for atoms in the F = 0 states, but it should be pointed out that it is feasible to construct the atomic beam apparatus in such a way that the F = 0 atoms are eliminated so that the proton polarization is maintained even in a weak magnetic field. The presence of a magnetic field in the ionizer often is desired for other reasons, e.g. to confine electrons. It is sometimes presumed that ionization has to take place in a strong field to avoid depolarization in the ionizing collision. This is not the case. It has been demonstrated that ionization of $\vec{H}^o$ by electron bombardment or by bombardment with fast atoms in a single collision does not affect the nuclear polarization, i.e, the expectation value of the proton spin is the same before and after the ionizing collision.

Below, typical values of beam polarization for different ionizers will be mentioned, but the processes which determine the attainable degree of polarization will not be discussed in detail. For a conventional atomic beam source, the polarization of the protons in the H^o atoms is expected to be ~ 96% for a magnetic field of 0.2T and ~ 92% for a field of 0.1T. These figures take into account that the atomic beam source accepts some atoms (1-2%) of the wrong electron spin ($m_j = -1/2$) and that the ionizer magnetic fields used are not strong enough to decouple electron and proton entirely (see ref. 1). In practice, proton polarization of 0.8 to 0.9 has been attained. The reduction from the expected values arises primarily from contamination of the ion beam with unpolarized beam from background gas.

2. IONIZATION BY ELECTRON BOMBARDMENT

The cross section for the process

$$\vec{H}^o + e \rightarrow \vec{H}^+ + 2e \qquad (1)$$

reaches a maximum of $\sigma \simeq 0.7 \times 10^{-16}$ cm^2 at an electron energy of ~ 60 eV (ref. 2). For higher energies σ decreases monotonically to 0.1×10^{-16} cm^2 at 1 keV.

If N hydrogen atoms are bombarded with electrons of current density j, the resulting $\vec{H}^+$ beam current $I(\vec{H}^+)$ is

$$I(H^+) = N \cdot j \cdot \sigma \qquad (2)$$

In a typical strong-field ionizer, the magnetic field is produced by a solenoid which is coaxial with the atomic beam. Electrons are injected into the solenoid from a filament near the solenoid axis. Positive ions are extracted from the opposite end of the solenoid by a field gradient which also serves to reflect the electrons and thus to confine them to the ionization region. The length of the ionization volume used to be typically 10 cm. Since in this case $N = 7 \times 10^{11}$, one expects

$$I(H^+) = 0.05\ \mu A \text{ per mA/cm}^2, \qquad (3)$$

provided the electron energy is at the cross section maximum. The best current obtained with this type of ionizer, first described by Glavish et al.[3] in 1966, was about 10 µA, which corresponds to an electron current density of 0.2 A/cm^2 and an ionization efficiency of 0.3%. The actual current density of course must be considerably larger since the electrons can not have the optimum energy throughout the ionization volume because of the gradients caused by space charge. In fact, the potential difference between the filament and the electrode inside the solenoid is always considerably larger than 60 eV [the cross section maximum] and is of the order 1 kV in modern ionizers.

During the last few years, the efficiency of electron bombardment ionizers has been improved by an order of magnitude. The new generation "super-ionizers" produce a DC current of about 100 μA $\vec{H}^+$ with a standard atomic beam source[4], which corresponds to an ionization efficiency of ~ 3%. The original development[5] was carried out jointly between ANAC and CERN. In these ionizers a plasma discharge is maintained in a solenoid magnetic field of moderate strength (~ 0.2T). The discharge is supported by electrons from a filament. A schematic cross section of the ionizer built by Grüebler's group at ETH, Zürich, is shown in fig. 1. The effective length of the

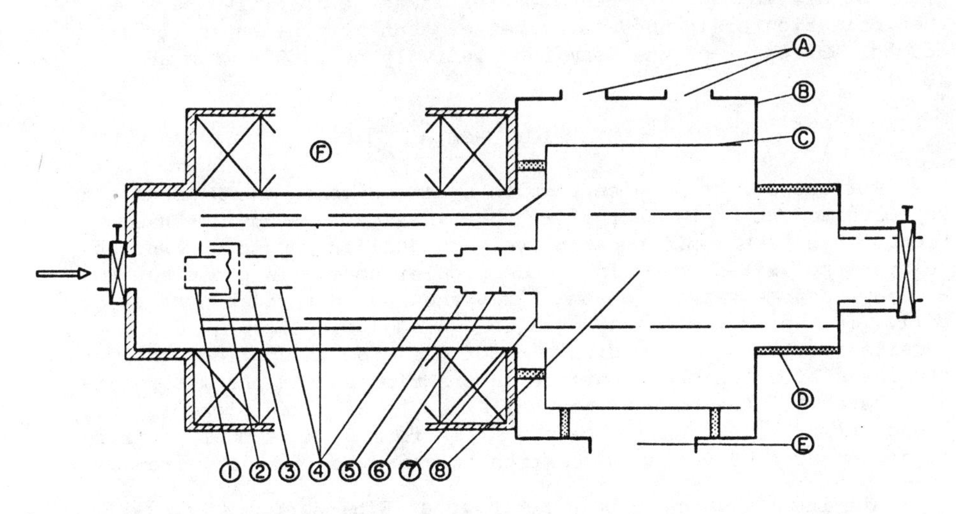

Fig. 1. Schematic of the ionizer: A, E: pumping ports, B: vacuum housing, C: electrode support, D: insulator, F: solenoid. Electrode system: 1: electron repelling, 2: filament and grid, 3: electron acceleration and ion repelling, 4: ionisation column potential, 5: electron reflexion and ion extraction, 6-8; acceleration, beam forming and transport (from ref. 4).

ionization region is 35 cm. The electron current density must be a few A/cm^2. Several commercial ionizers[6] of this type are now in routine operation. Note that at the Argonne ZGS, about the same beam intensity (100 μA) was also obtained with the more conventional, shorter ionizer by pulsing the RF discharge and by cooling the atomic beam[7]. If these techniques are combined with a super-ionizer one should expect pulsed polarized beams approaching 0.5 mA. A beam polarization of P = 0.85 and an emittance of $\lesssim 12\pi$ mm mrad (MeV)$^{\frac{1}{2}}$

can be expected[4]. The above estimate of the emittance is based on the beam transmission characteristics observed in the ETH tandem.

It is not known in detail what limits the ultimate ionization efficiency of current ionizers. Possibly, something can still be gained by increasing the length of the ionization volume, although extraction of the ions will at some point become difficult. Presumably even larger electron current densities can be confined if the solenoid magnetic field is increased. This approach is being studied at Bonn[8] and at Saclay[9], employing superconducting solenoids. Recent tests at Bonn with a Penning discharge in a 4T magnetic field show that ions can be extracted only from a very limited radial region in the ionizer (~ 3 mm diameter) because for ions created too far from the axis the solenoid fringe field imparts too large a transverse velocity to the ions. At the present stage of development the Bonn ionizer is not competitive because the ionization efficiency is no better than for the super-ionizer and the emittance of the beam is relatively poor [90π mm mrad $(MeV)^{\frac{1}{2}}$].

3. CONVERSION FROM $\vec{H}^+$ TO $\vec{H}^-$

For accelerators which require negative ions, the $\vec{H}^+$ beam can be converted to $\vec{H}^-$ by charge exchange in a gas or vapor. This process has been applied extensively by nuclear physicists who accelerate polarized beams in tandem accelerators. In practice, the $\vec{H}^+$ beam is accelerated to some 5 keV and passed through a Na vapor cell. In the charge exchange region, a magnetic field (~ 0.2T) parallel to the momentum direction of the beam is applied, to avoid depolarization via the hyperfine interaction during the intermediate state when the beam is neutral, i.e., the $\vec{H}^+$ has picked up one (unpolarized) electron. Once the $\vec{H}^-$ is formed, a decoupling field is no longer needed since H^- exists only in the (singlet) ground state.

The best beams have been achieved at ETH, Zürich where 3 μA $\vec{H}^-$ are produced by charge exchange of 100 μA $\vec{H}^+$ at 5 keV energy in Na vapor. Separate measurements of the equilibrium fraction yield $F_-^\infty = 0.07$ for 5 keV H^+ in Na (see fig. 2) or about twice the value obtained in operating $\vec{H}^-$ polarized ion sources. The difference is thought to arise from the insufficient ion-optic acceptance of the Na charge exchange canal presently in use. There is good reason to believe that this problem can be solved without leading to unreasonable difficulties from increased flow of Na vapor out of a large diameter charge exchange canal. If this improvement is added to gains resulting from pulsing the dissociator and cooling the atomic beam, pulsed $\vec{H}^-$ currents of 20 μA should be possible in routine operation. Schmelzbach et al.[3] anticipate that $\vec{H}^-$ beams in excess of 100 μA can be achieved. Experience at ETH indicates that conversion to negative ions can be accomplished without substantial loss in polarization or emittance.

Larger equilibrium yields, F_-^∞, are observed at lower energies, in particular for charge exchange in Cs and Rb (fig. 2). In

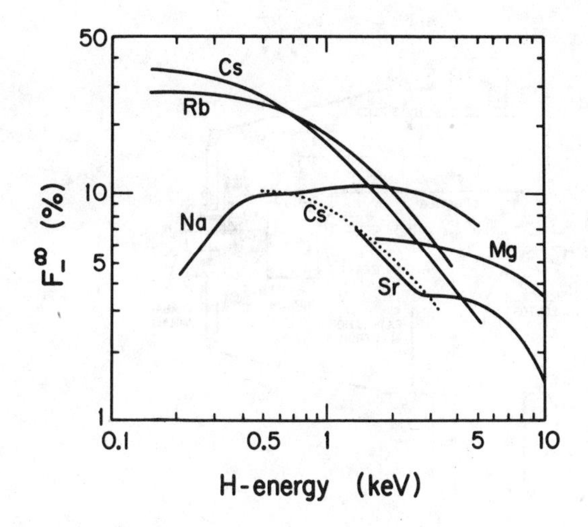

Fig. 2. Equilibrium negative ion yield in alkali vapors (from ref. 10). The dotted line for Cs is from ref. 11.

present sources, a beam energy of 5 keV (and thus Na as a charge exchange medium) is used to avoid the serious ion optic problems associated with the design of a charge exchange system for a beam of 1 keV or below. It would be useful to discuss the possible use of Cs or Rb as a charge exchange vapor at this workshop. It should be mentioned, however, that there still are considerable uncertainties in the values of F_-^∞. To illustrate this point, two relatively recent measurements for Cs are shown in fig. 2. The difficulties of measuring F_-^∞ at low energies probably is related to loss of beam by scattering in the charge exchange cell. Loss of beam by scattering makes it questionable whether the large values of F_-^∞ obtained under ideal circumstances can be realized in practical application to polarized H^- ion sources.

4. DIRECT CONVERSION OF $\vec{H}^o$ TO H^- BY THE COLLIDING BEAM METHOD

The colliding beam principle developed at Wisconsin[12,13] has the unique feature of converting polarized thermal $\vec{H}^o$ atoms directly into $\vec{H}^-$ without going through positive ions as an intermediate step. The idea is to bombard the H^o atomic beam with a beam of fast atoms or ions which act as the donor of electrons. Of the two reactions which have been proposed:

$$\vec{H}^o + Cs^o \rightarrow \vec{H}^- + Cs^+ \tag{4}$$

and

$$\vec{H}^o + D^- \rightarrow \vec{H}^- + D^o \tag{5}$$

the first is used in the source installed on the Wisconsin tandem accelerator[23]. The ionizer of this source is shown in fig. 3. The atomic beam enters from the left. Ions created in the collision region are extracted towards the right and are deflected by an electrostatic spherical deflector. The Cs^o beam is collinear

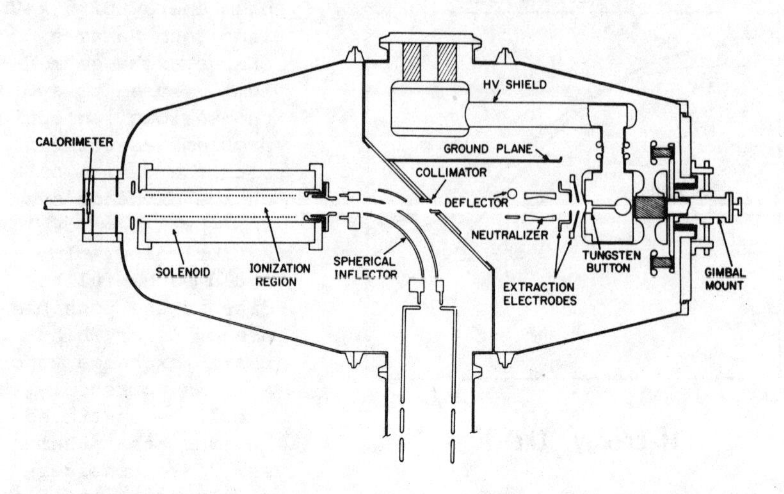

Fig. 3. Colliding-beams source for polarized negative ions. The atomic beam enters from the left. The exit port of the polarized negative ions is on the bottom of the drawing. The atomic-beam source is not shown. For normal operation, the calorimeter, which serves only to adjust the Cs-gun, swings out of the way.

with the H^o atomic beam. Cesium is ionized on a porous tungsten disk of 3 cm^2 area, heated to some 1200°C. The Cs^+ gun and electrode system is gimbel-mounted such that the Cs beam can be aimed along the atomic beam axis. The 40 keV Cs^+ beam is neutralized in Cs vapor. The source, which is described in more detail in ref. 14, produces beam currents up to 3.3 μA. The proton beam polarization is very high [P = 0.91±0.01 measured after acceleration, for ionization in a 0.1T field], because the cross section for production of H^- by collision of Cs^o with H_2 and hydrocarbon molecules is much smaller than for H^o. The emittance of the beam is measured[14] to be below 10π mm mrad $(MeV)^{\frac{1}{2}}$.

It is interesting to note that the observed beam intensity is in quantitative agreement with a calculation, based on the known charge exchange cross section, and the known atomic beam density and Cs^o current density. The observed beam current used to be somewhat higher than expected[13] but this is explained by newer cross section measurements[10] which give larger values than earlier ones. If we assume $N = 25 \times 10^{11}$ atoms in the collision region (35 cm length) and a cross section of 6×10^{-16} cm^2 for reaction (1), the beam current expected according to eq. (2) is

$$I = 1.5\ \mu A\ \vec{H}^- \text{ per mA/cm}^2\ Cs^o. \quad (6)$$

This is in excellent agreement with the observed value of 1.3 μA per particle-mA/cm^2 Cs^o. These numbers refer to a conventional atomic-beam source, and do not include the potential gain from a cooled or pulsed dissociator.

Substantial gains in beam intensity can be expected in the future since, compared to other sources, little development work has been done. The present limitation to 3 μA is related to the poor focus of the Cs beam, which causes excessive loading of some of the ionizer power supplies. To study this problem, a separate test setup for the Cs gun has recently been constructed. The results are encouraging, in that 8 particle-mA of 55 keV Cs^o were observed in the calorimeter (1 cm diameter aperture), compared to ~ 2 mA in the present source, with only a 50% increase in total Cs$^+$ output of the source. With the new Cs gun, a cooled atomic beam and a pulsed dissociator, beam currents of 20 μA should be achieved in routine operation.

It would be very interesting also to develop a source based on D$^-$ instead of Cs^o as the electron donor. The charge exchange cross section in this case is resonant and increases like 1/v. For a D$^-$ energy of 2 keV the cross section for reaction (4) is 20×10^{-16} cm^2. If we again assume a standard atomic beam source and a collision region of 35 cm length, we expect

$$I = 5\ \mu A\ \vec{H}^- \text{ per mA/cm}^2\ D^- . \tag{7}$$

Very intense sources of D$^-$ have been developed for the CTR program. They have sufficiently good emittance to permit deceleration to 2 keV. A serious problem with this type of reaction is the loss of the newly-formed $\vec{H}^-$ by the space charge of the D$^-$ beam, but several methods to overcome the problem have been proposed[5,15].

The production of *positive* ions by the colliding-beam method, e.g.

$$\vec{H}^o + \vec{H}^+ \rightarrow \vec{H}^+ + \vec{H}^o \tag{8}$$

has been proposed some time ago[16]. It is unlikely that a ionizer based on this reaction would be competitive with the new high-efficiency electron bombardment ionizers.

5. SURFACE IONIZATION

It may be of interest to mention surface ionization as a possible method to produce H$^-$ ions directly from thermal $\vec{H}^o$ atoms. *Negative* ions will be formed when an atom strikes a surface whose work function is comparable to the electron affinity of the atom. Since the electron affinity of hydrogen atoms is 0.75 eV, a surface of very low work function is required to obtain appreciable H$^-$ by surface ionization. A number of tests have been reported[17,18]. As far as I know the highest value of F_- observed so far ($F_- \sim 10^{-5}$) was obtained with a tantalum foil coated with a barium/strontium oxide mixture[18] (surface temperature 1200 K). An efficiency $F_- \sim 10^{-3}$ is required to produce 3 μA H^- by ionization of H^o atoms from a standard atomic beam source.

6. CONCLUSIONS

The best electron bombardment ionizers currently produce polarized DC beams of 100 μA H^+ and D^+, with good emittance and high polarization. Further substantial gains in beam intensity can be expected from improvements in the atomic beam apparatus. The work shop should address the question of further developments of the ionizer, even if the expected gain in ionization efficiency are not dramatic. It should be pointed out once more that an improvement by a factor 2 should be considered a highly significant result of a development project, even though it lacks the glamour of clever new proposals which propose to improve beams by orders of magnitude.

Charge exchange of $\vec{H}^+$ to $\vec{H}^-$ has produced 3 μA of polarized negative ions. Here, an improvement of a factor 2 or so can be expected from increasing the ion-optic acceptance of the charge exchange cell, in addition to the possible improvements in the $\vec{H}^+$ beam. The workshop might discuss whether there are additional opportunities for improvements from charge exchange in Rb or Cs at lower energies.

Bombardment of the atomic beam with a collinear beam of 40 keV Cs^o has yielded 3 μA $\vec{H}^-$ and $\vec{D}^-$. A more intense Cs^o gun shows promise for a factor four improvement. Since this ionization method has seen little development so far, the workshop might consider new conceptual designs of Cs^o guns. These need not necessarily be limited to collimated, linear beams of Cs^o, but could include other geometries (bombardment with transverse or radial beams). Also one might ask whether gains can be expected from pulsing of the Cs gun. Of particular interest would be a discussion group to develop a practical design for ionization of $\vec{H}^o$ by bombardment with D^-, since in principle one sees an opportunity for very large $\vec{H}^-$ intensities.

In view of the limited effort that has gone into the development of polarized ion sources during the last few years, progress in the design of new ionizers has been impressive. There is reason to believe that a systematic, intensive development program will bring pulsed $\vec{H}^+$ beam currents of 1 mA and pulsed $\vec{H}^-$ currents of 100 μA within reach.

This work was supported in part by the U.S. Department of Energy.

REFERENCES

1. W. Haeberli, Ann. Rev. Nucl. Sci. 17, 373 (1967).
2. L.J. Kieffer and G.H. Dunn, Rev. Mod. Phys. 38, 1 (1966).
3. H.F. Glavish, E.R. Collins, B.A. McKinnon and I.J. Walker, Proc. 2nd Symp. Polarization Phenomena (Karlsruhe), Experientia Sup-

plement 12, 85 (1966).
4. P.A. Schmelzbach, W. Grüebler, V. König and B. Jenny, Nucl. Instr. Meth. 186, 655 (1981).
5. H.F. Glavish, Higher Energy Polarized Proton Beams (Ann Arbor, 1977) AIP Conf. Proc. No. 42, p. 47; W. Kubischka, CERN, private communication.
6. ANAC, Inc., 3067 Olcutt St., Santa Clara, CA. 95050, U.S.A.
7. P.F. Schultz, E.F. Parker and J.J. Madsen, AIP Conf. Proc. No. 69, p. 909.
8. A. Kruger, H.G. Mathews, S. Penselin and A. Weinig, Nucl. Instr. Meth. 138, 201 (1976); and A. Kruger, Ph.D. thesis, University of Bonn (1979).
9. R.M. Beurtey, AIP Conf. Proc., No. 51, p. 330 (1979).
10. A.S. Schlachter, K.R. Stalder and J.W. Stearns, Phys. Rev. A 22, 2494 (1980).
11. C. Cisneros, I. Alvarez, C.F. Barnett and J.A. Ray, Phys. Rev. A14, 76 (1976).
12. W. Haeberli, Nucl. Instr. Meth. 62, 355 (1968).
13. D. Hennies, R.S. Raymond, L.W. Anderson, W. Haeberli and H.F. Glavish, Phys. Rev. Lett. 40, 1234 (1978).
14. W. Haeberli, M.D. Barker, C.A. Gossett, D.G. Mavis, P.A. Quin, J. Sowinski and T. Wise, Nucl. Instr. Meth. (to be published).
15. W. Haeberli, High Energy Physics with Polarized Beams and Polarized Targets, Argonne, 1978. AIP Conf. Proc. No. 51, p. 269.
16. R. Beurtey and M. Borghini, J. Phys. C2, 56 (1969).
17. W.G. Graham, Bull. Am. Phys. Soc. 23, 409 (1977); R.S. Raymond, W. Haeberli and P.A. Quin, Nucl. Phys. Progress Report (1978), (unpublished).
18. R. Goldstein and J.E. Graf, JPL Invention Report 30-3834/NPO-14113, Cal. Tech., Pasadena, California.

PROGRESS ON IONIZATION STAGE

S. Jaccard
Swiss Institute for Nuclear Research, CH-5234 Villigen

INTRODUCTION

This paper is an attempt to report on the high lights of the four "Ionization Working group" meetings. The theme primarily was the ionization of polarized thermal $\vec{H}^{0}$ atoms, with a special emphasis on the production of $\vec{H}^{-}$.

ELECTRON BOMBARDMENT $\vec{H}^{0} + e \longrightarrow \vec{H}^{+} + 2e$

The successful operation of the new generation of strong field electron bombardment ionizers, the so-called "super ionizers" [1,2], raised the question of further improvements. It has been pointed out that the electron current density in those ionizers is already quite impressive. In his review talk, Haeberli showed that with 10 cm of interaction region, the current of $\vec{H}^{+}$ to be expected from a $10^{11}\,\vec{H}^{0}/cm^{3}$ atomic beam is given by:

$$I = 0.05\ \mu A \text{ per } mA/cm^{2}\ e^{-}$$

if it is assumed that the electron energy is near the cross section maximum (~ 100 eV). Currents of 100 μA of $\vec{H}^{+}$ are obtained with 30 cm long ionizers with half this atomic beam intensity. Therefore the electron density in the ionizer is in excess of 1 A/cm^{2}.

It was the general agreement of the participants that little is really understood of the conditions yielding these high efficient modes of operation. In a letter to prof. Krisch, G.J. Witteveen wrote however: "The steep increase of the ionization efficiency can be fully explained by the calculations of Smith and Hartmann". It was a surprise to find that these calculations were done in 1940 [3]. Reference was made to the similarity of conditions met in the space charge lenses developed by the very high beam current community[4].

A direct observation of the life time of the electrons inside the ionizer is worth mentioning here. If the ionizer operates without filament current, a proton beam intensity increase of about 1 % is observed when the electrons of the incoming atomic beam are changed from a polarized to an unpolarized state (switching on of one HF cavity only). The origin as well as the use of this effect for tuning the high frequency cavities of a polarized source at injection energy has been demonstrated elsewhere[5]. Table I gives the spin states of electron and proton for the four energy levels of the hydrogen atom in a strong magnetic field.

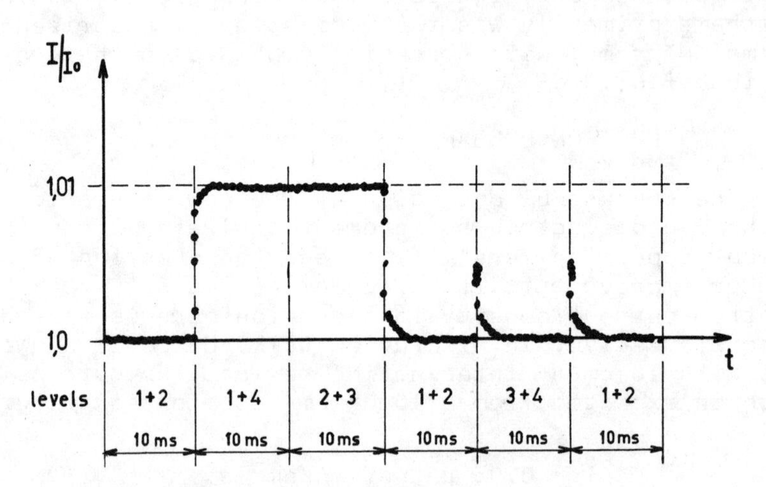

Fig. 1. Beam intensity changes observed as atomic levels are selected by the RF cavities.

Table I Labelling for the four split levels of the ground state atomic hydrogen in a strong magnetic field.

level number	1	2	3	4
m_j	+1/2	+1/2	-1/2	-1/2
m_I	+1/2	-1/2	-1/2	+1/2

Fig. 1. shows the current out of the ionizer as seen on a signal averager scope when the RF transitions are sequentially put in operation. As in this mode of operation the electrons trapped in the ionizer are mainly

electrons stripped from the hydrogen atoms themselves, the little peaks observed as one switches from a given electronic polarization to the reversed one are a signature of the life time of the electrons in the ionizer.

The question was raised of the possibility to increase still further the length of such ionizer. There is no evidence that this would lead to a higher beam emittance.

CHARGE EXCHANGE $\vec{H}^+ + X^o \longrightarrow \vec{H}^- + \ldots$

Gruebler reported on the 3 μA of $\vec{H}^-$ produced by charge exchange in a sodium cell with the 100 μA, 5 keV $\vec{H}^+$ beam of the ETH atomic beam source. It is his belief that technical improvements in the cell design will lead to 6 % charge exchange efficiency.

A cesium cell, with a beam of one keV or less should lead to a 2 to 2.5 higher efficiency than a sodium cell if used at lower incident energy.

Steffens pointed out that one should not design too short a charge exchange cell, as the process happened mainly in two steps, some having long intermediate decay, like the $H^o(2s)$ which contributes substantially to the total amount of H^- produced.

A proposition from Sluyters was briefly looked at. It is to use the up to 30 % efficient charge exchange reaction of a grazing H^+ beam on a cesium coated surface. The required deceleration of the beam to less than 100 eV might be inconsistent with the rather stringent requirements for grazing incidence.

For the sake of comparison with other schemes discussed further, we summarize in table II the expectations for beam intensities out of a polarized atomic beam source. We take for granted that a factor of two improvement on the actual performances of the electron bombardment ionizer is a fair estimate. $10^{11}\vec{H}^o/cm^3$ is the beam density produced by a STANDARD atomic source operating at room temperature. We take $10^{12}\vec{H}^o/cm^3$ as what should be expected from a COLD atomic source, in which the dissociator operates at liquid nitrogen temperature and the nozzle is cooled at 4^oK. According to what has been discussed at this workshop, $2 \cdot 10^{13}\vec{H}^o/cm^3$ is what might be expected from a Kleppner type of ULTRA COLD atomic source.

Table II: Beam expectations from an improved electron bombardment atomic beam source.

Type of atomic source	$\vec{H}^+$ current (mA)	$\vec{H}^-$ current with Na cell	$\vec{H}^-$ current with Cs cell
Standard	.2	12 μA	30 μA
Cold	2	120 μA	300 μA
Ultra cold	40	2.4 mA	6 mA

Needless to say, the figures presented in table II are optimistic. However, no serious drawback effect is expected in the ionizer, as even in the ultra cold atomic source case the ion current remains less than 5 % of the electron current and ionization probability is less than 13 %. Operation of charge exchange cells with such **a high intensity** H^- beam might be a problem, but **operating it in a pulsed mode looks very** promising.

ECR IONIZER ?

One interesting feature of an ECR source is the high density (10^{12}e/cm^3) of hot electrons achieved[6]. This should be compared with the 1 A/cm^2 mentioned earlier, i.e. with less than 10^9e/cm^3 confined in an electron bombardment ionizer. Although the energy of at least 1 keV in a ECR source is somewhat too high for optimum ionization cross section, it seems that very efficient ionizers could be designed that way. Concerns were expressed about the extraction from a somewhat higher magnetic field (0.3 T to be compared with 0.15 T) and also about possible depolarization effects. One should look further into this scheme.

LASER IONIZER ?

Few words were spoken about the possibility of using laser beams for ionization purpose. Steffens shortly reported on the 10 pA of beam produced at Bielefeld.

Competition with other ionizers was ruled out because of the short wave length very high laser power needed.

ION BOMBARDMENT $\vec{H}^{o} + X^{+} \longrightarrow \vec{H}^{+} + X^{o}$

Not very many thoughts were given to such a scheme, probably because of the already good results obtained with the super ionizers. However Arvieux mentioned that there is reason to believe that good results might be obtained from a $\vec{H}^{o} + H^{+} \longrightarrow \vec{H}^{+} + H^{o}$ ionizer operating in space charge neutralized condition.

COLLIDING BEAM METHODS

This method is successfully in operation at Madison. 3 μA of $\vec{H}^{-}$ are obtained using the $\vec{H}^{o} + Cs \longrightarrow \vec{H}^{-} + Cs^{+}$ charge exchange reaction at about 40 keV. Haeberli reported in his talk on an improved version of the Cs gun which deliver four time as much beam, namely up to 10 mA/cm² of Cs^{o}. It is believed that technical improvements should at least double that figure.

The $\vec{H}^{o}$ atomic beam source at Madison is of a standard type. As seen earlier in this talk, cryogenic technology should improve the beam density by one or two order of magnitude. Quin mentioned some difficulties to be expected. The life time of the dissociator will be drastically reduced when the Cs beam intensity is increased due to Cs deposit in the nozzle. Some changes in the geometry are therefore required. Cesium beam halo in the neutralizer has to be kept small. All of these drawbacks are strongly reduced in a pulsed mode of operation. Table III gives the $\vec{H}^{-}$ beam intensities "blindly" expected with improved $\vec{H}^{o}$ and Cs beams and for 30 cm inter-action region.

Table III: Beam expectations from an improved Cs colliding beam source.

Type of $\vec{H}^{o}$ atomic source	Standard	Cold	Ultra cold
$\vec{H}^{-}$ current	25 μA	250 μA	5 mA

Fig. 2 shows the conceptual design by Moffet of an interesting alternative to a colliding beam source with ultra cold $\vec{H}^0$ beam. A high intensity pulse of 100 keV Cs atoms is carefully focused on the charge exchange region of a 4.8 T ionizer. The Cs atoms collide

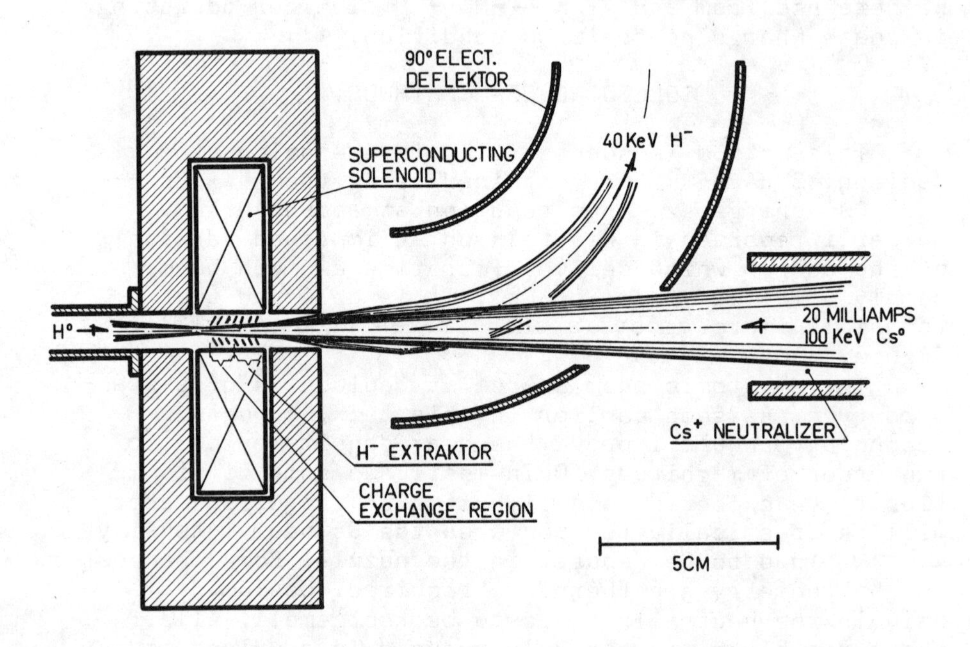

Fig. 2. The Moffet conceptual design of a 5 mA pulsed polarized H^- source.

with a high density of $\vec{H}^0$ atoms obtained by cooling down an atomic beam on electrodes at typically 0.5°K (He_3 coating). Pierce-type electrodes should extract at 40 keV a pulse of $\vec{H}^-$ of up to 5 mA.

Other candidates for the production of an intense $\vec{H}^-$ beam are the reactions

$$\vec{H}^0 + D^- \longrightarrow \vec{H}^- + D^0 \text{ and } \vec{H}^0 + H^- \longrightarrow \vec{H}^- + H^0$$

where the deuterons (protons) have an energy of 2 keV (1 keV) or less. Table IV gives the currents expected with a 12 mA/cm^2 D^- beam[7] of 1 keV with 20 cm of inter-action region.

Table IV: Beam expectations from a $\vec{H}^0 + D^- \longrightarrow \vec{H}^- + D^0$ colliding beam source.

Type of $\vec{H}^0$ atomic source	Standard	Cold	Ultra cold
$\vec{H}^-$ current	50 μA	500 μA	10 mA

Here the obvious and serious problem is the space charge blow up. Several schemes have been discussed in which reasonable space charge neutralization should be achieved. Fig. 3. illustrates such a scheme. A few other

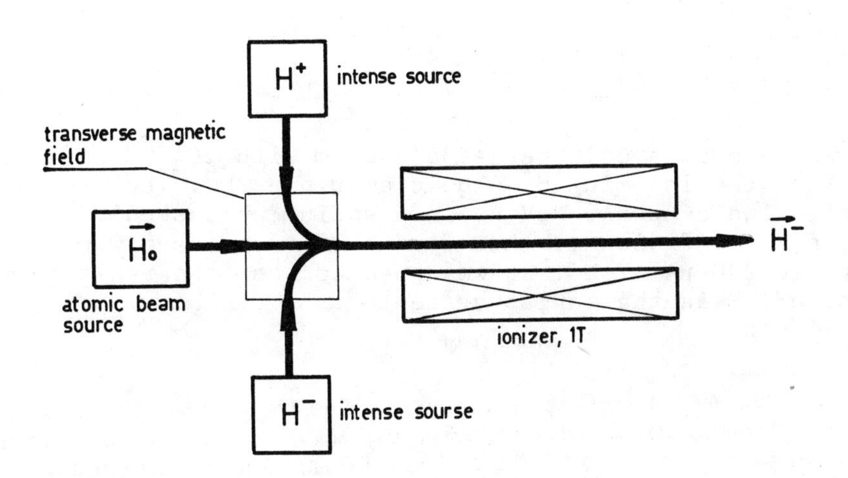

Fig. 3. Conceptual design of a space charge neutralized colliding beam source.

ideas can be found in the minutes of the High Intensity Unpolarized Source Working Group of Thursday, May 21.

SURFACE IONIZATION

Haeberli pointed out the possible application of surface ionization to the direct conversion of $\vec{H}^0$ atomic beam to $\vec{H}^-$. Since the electron affinity of the hydrogen atom is 0.75 eV, a surface of very low work function is required to obtain an appreciable yield. A number of tests have been reported[8]. The highest ionization efficiency reported so far is only about 10^{-5}. Anderson suggested that a surface of very low work function might be produced in a semi-conductor surface in which the conduction band would be enriched thermally or optically. What can be achieved this way is an open question.

IONIZATION OF FAST $\vec{H}^0$

A peculiar scheme for a polarized source emerged from the Round Table Discussion of May 24.

$$H^+ + \vec{D}^0 \longrightarrow \vec{H}^0 + D^+$$

$$\underset{5\ \text{keV}}{|} \qquad \underset{5\ \text{keV}}{\llcorner} \longrightarrow \vec{H}^0 + Na^0 \longrightarrow \vec{H}^- + Na^+$$

If we suppose a cold target of 20 cm with $10^{13}\vec{D}^0/\text{cm}^3$, 40 % of the incoming H^+ should emerge as $\vec{H}^0$. The ionization of the 5 keV $\vec{H}^0$ in a sodium cell should reach 7 % efficiency. Therefore a 10 mA H^+ beam would produce 200 μA of $\vec{H}^-$. It was checked that the replacement of $\vec{D}^0$ in the target presents no significant difficulty.

REFERENCES

1. P.A. Schmelzbach et al., Nucl. Instr. Meth. 186 655 (1981) and Proceedings of the 1980 Int. Symp. on High Energy Physics with Polarized Beams and Polarized Targets, 38 Experientia Supplementum, p.432.
2. H.F. Glavish, IEEE Trans. Nucl. Sci. NS-26, 1517 (1979).
3. Smith and Hartmann, J. Appl. Phys. 11, 220 (1940).

4. R.M. Mobley et al., H.W. Lefevre and R. Booth, IEEE Trans. Nucl. Sci. NS-26, No. 3, 3112-3117 (1979).
5. S. Jaccard, Proceedings of the 1980 Int. Symp. on High-Energy Physics with Polarized Beams and Polarized Targets, 38 Experientia Supplementum, p.443.
6. V. Bechtold et al., Nucl. Instr. Meth. 178, 305 (1980).
7. M. Delaunay et al., 2nd Int. Symp. D^- sources, Brookhaven, Oct. 1980.
8. W. Haeberli, Proceedings of the 1980 Int. Symp. on High-Energy Physics with Polarized Beams and Polarized Targets, 38 Experientia Supplementum, p.199.

Report on the Ionization Stage Work Session I

Thursday, May 21, 1981

Chairman: W. Haeberli Secretary: S.R. Magill

The following topics were suggested for discussions: 1) Electron bombardment; 2) Ion and atom bombardment; and 3) Exotic ionization methods including surface ionization, photons, and plasmas.

1. Electron bombardment - Discussion was concentrated on an attempt to understand the electron bombardment ionization method by comparison of various ionizers. It was concluded that these devices are not well understood. At ETH, when the useful effective length was increased from ~7 cm for the old ionizer to 30 cm for the new ionizer, the ionization efficiency increased by a factor of 10. A possible explanation was that the increased length produced an effective potential that trapped the electrons better, thereby providing more efficient ionization. Experience with the new ANAC ionizer at Saclay indicated that mechanical alignment and stability are important. It was concluded that useful improvements in the ionization efficiency and therefore beam intensity from electron bombardment ionizers can be expected from the results of further experimentation with longer ionizers. Also, the effect on the beam from changes in electrode geometry should be studied. Development of good pulsed operation is important if one plans to store the beam in a storage ring prior to injection into the main accelerator.

2. Ion and atom bombardment ionizers were mentioned briefly but referred to later working groups.

3. Exotic ionization methods - L.W. Anderson suggested that surface ionization of H° to H^- might benefit from a semiconductor surface where electrons are raised to the conduction band by photons or heat. These electrons would then have a sufficiently low work function to ionize H° atoms striking the surface. A suggestion involving $H^\circ \rightarrow H^-$ by grazing incidence surface ionization was referred to a later time.

The use of photons (laser) as an ionizer was also referred to a later time.

A method was suggested to use electron cyclotron resonance (ECR) to strip electrons from H°. The following reference was given:

M. Delaunay, et al.,; Association EURATOM-CEA, Grenoble; Proceedings of 2nd International Symposium on Production and Neutralization of H^- Ions and Beams, Brookhaven (BNL 51304), 1980, pg. 255.

Ionization efficiencies by this method are probably high, but depolarization could be a problem. The ECR method will also be investigated further in future working sessions.

Report on the Ionization Stage Work Session II

Friday, May 22, 1981

Chairman: W. Haeberli Secretary: S.R. Magill

The discussion started with P. Schultz of Argonne further explaining the development and use of alumino-silicate Cs ion emitters. These emitters are made by mixing powders of Al_2O_3, Cs_2CO_3, and 2 parts SiO_2. This mixture is then heated in an oven to ~1400°C, evolving CO_2 and forming an alumino-silicate lattice with Cs^+ ions trapped inside. Pellets of this material are placed in a commercially available heating element and are then ready for use. Cs^+ ions are drawn out of the pellet by heating it to ~1000°C and applying an accelerating voltage. The lifetime is reported to be long and this type of emitter seems to be especially attractive for pulsed beam operation due to elimination of the tungsten surface ionization emitter.

The problem of Cs deposition in other regions of the source (e.g. atomic beam stage) is manageable and does not appear to be a serious problem.

W. Haeberli described the recirculation type Cs neutralizer used in the Wisconsin source. The sloped walls of the chamber provide a means for Cs vapor to condense (at 35°C) and flow back into the reservoir. This neutralizer operates with 85-90% neutralization efficiency and has a Cs consumption rate of ~ 2.5 g/day.

A question was raised about the choice of RF frequency used for dissociation of H_2 in the atomic beam stage. Is dissociation more efficient using RF in the range 100 MHz → 1GHz? The experience of most of the participants indicated that there is no evidence for needing higher frequency dissociation.

Report on the Ionization Stage Work Session III

Tuesday, May 26, 1981

Chairman: S. Jaccard   Secretary: S.R. Magill

In this last work session on the ionization stage, the discussion was a review of the topics addressed in previous sessions. An attempt was made to try to evaluate each of the methods discussed as to its advantages, disadvantages, improvements, and chances of succeeding. The results of these discussions will be presented in the Ionization Stage Progress Report.

Problems of depolarization were discussed for the surface ionization semiconductor scheme and for the ECR ionizer. It was felt that this should be investigated before evaluation could be made. Laser stripping ionization was generally ruled out because of the large amounts of power needed. The colliding H°, Cs beam method was seen as the best method for short term improvement. Space charge problems seem to begin only at H^- currents somewhat over 100 μA. Extraction of H^- was seen to be still a problem in the H°, D^- scheme. The cross-section improvement of ~ factor of 10 may, in fact, disappear when other problems are encountered. Witteveen's letter was discussed in regard to his remarks about the shaping of magnetic extraction fields, and suggestion of his spin-state selection by B-field shaping.

The discussion was concluded with R. Moffett showing a proposal for a 5mA H^- source using the H°, Cs colliding beam method. Features of this method are a 48 kG ionizer with Pierce-type extraction electrodes and high density H° injection. A diagram of this method will be given in the Progress Report.

Report on the $H^+ \rightarrow H^-$ Ionization Work Session

Monday, May 25, 1981

Chairman: W. Gruebler Secretary: S.R. Magill

The desirability for $\vec{H}^-$ use in High Energy accelerators has led to the development of several schemes which have been discussed in this workshop. The subject of this working group is the double charge exchange reaction between H^+ and certain alkali metals producing H^-.

This reaction was described in the following reference:

S.K. Allison, Rev. Mod. Phys. 30, 1137 (1958).

Cross sections for various alkali metals vs. incident H^+ energy were given in:

W. Gruebler, et al., Helvitica Physica Acta, Vol. 43, 1970.

The charge exchanger consists of an alkali vapor canal which serves as a target for the H^+ beam. One attractive combination is that of a 5 keV H^+ beam incident on Na vapor due to its large charge exchange cross-section. Also, the straggling angle (angular spread of beam by charge exchange process) for this reaction is small ($\sim 0.5°$).

An area in which there is considerable room for improvement is in the total $H^+ \rightarrow H^-$ charge exchange efficiency. Currently, for H^+ through Na vapor, 100 μA H^+ yields ~ 3 μA H^-. A large part of the problem is in transporting the 100 μA H^+ beam through the Na vapor canal. A typical canal size is 1 cm diameter and 12 cm long. For this geometry, with no Na present, 100 μA H^+ incident beam is transported through the tube with ~ 33% efficiency yielding 33 μA H^+. Addition of Na (5×10^{-3} torr-cm target thickness) yields the total $H^+ \rightarrow H^-$ efficiency of ~3%. It was suggested that shortening the canal would result in higher transmission efficiency with the target thickness provided by a Na vapor jet. A sheet of Na vapor several cm wide and a few mm thick was used at CERN for beam profile measurements. A target this thin could not be used for the $H^+ \rightarrow H^-$ charge exchange, however, because formation of the H^- involves 3 steps, one of which involves a relaxation time for H° to change from an excited state to the ground state. This relaxation time makes the minimum length of the vapor canal at least several cm. This should be calculated in more detail. Another suggestion was to replace Na with Cs and operate at low (~ 1 keV) H^+ beam energy. This method, however, would probably result in worse beam scattering problems than are currently encountered in Lamb shift sources.

In summary, it was generally agreed that the 3% yield currently obtained could be increased to ~ 7% in present designs. It was also thought that the use of jet targets could decrease vapor canal length resulting in more efficient beam transport. Another improvement would be to replace Na with Cs or Rb and operating at lower incident beam energies. Implementation of all these improvements could possibly result in a factor of 2-3 improvement in H^- yield.

COLD HYDROGEN TECHNIQUES FOR POLARIZED PROTON PRODUCTION

Daniel Kleppner
Department of Physics
Massachusetts Institute of Technology
Cambridge, Massachusetts, 02139

ABSTRACT

Methods have been developed for producing atomic hydrogen in the liquid helium temperature range, and also below 0.5K. These low temperature techniques provide opportunities for substantial improvements in polarized proton ion sources and related devices. Among the most dramatic of the new techniques is a simple method for producing a dense gas of atomic hydrogen with high nuclear polarization. This paper summarizes recent progress, including developments which have occurred since the PPIS Workshop, and describes some possible applications.

I. INTRODUCTION

New techniques for working with atomic hydrogen at low temperatures have emerged during the past few years from the quest to produce (electron) spin-polarized hydrogen. The goal of this research is observation of the Bose-Einstein transition, a phase transition which occurs when the deBroglie wavelength becomes comparable to the interparticle separation[1]. At 0.3K the transition in hydrogen requires a density of $8\times10^{19}/cm^3$. Although such a density has not yet been achieved, the techniques developed to produce spin-polarized hydrogen appear to have useful applications to many other lines of research, particularly to the production of polarized proton ion sources and jets, and possibly to polarized proton targets. This paper reviews these techniques and presents results on the creation of a proton-polarized gas. There are many ways to implement and adapt cold hydrogen techniques to polarized proton research; the ideas put forward here represent essentially a "first look" at these possibilities rather than a methodical study. It is hoped that they will be useful in illustrating some of the new opportunities and in pointing the way to useful designs.

II. BACKGROUND

By "cold hydrogen" techniques we refer to methods for dealing with atomic hydrogen at temperatures below, for instance, 8K. The subject naturally divides itself into what we might call the liquid helium region, 4K to 8K, where frozen H_2 can be used as a surface coating and conventional cryogenic techniques can be employed, and a very low temperature region, 0.2K to 0.4K, where liquid helium surfaces must be used and a dilution refrigerator or ^{3}He refrigerator is required. Production of spin-polarized hydrogen requires working in the very low temperature region. It is important, however, not to confuse the general subject of cold hydrogen technology with the specific techniques for producing spin-polarized hydrogen - the combination of very low temperature, high magnetic field and exotic wall covering needed to achieve stable samples of gas at relatively high density. Many applications require only the helium temperature region where none of these complications are involved.

The production of proton-polarized atomic hydrogen at unprecedented densities in the low temperature regime offers important new opportunities for polarized proton physics, and it is in this regime that one can expect the most dramatic advances. Nevertheless, the helium temperature regime also offers opportunities for significant improvements which deserve consideration. In Section III we discuss operation of a hydrogen atomic beam source in the helium temperature regime. In Section IV we describe the creation of a proton-polarized hydrogen gas in the low temperature regime, and its applications to polarized ion production.

III. 4K - 8K REGION

IIIA. General Description

Atomic hydrogen can be effectively thermalized by collisions with a surface coated with frozen H_2 over a temperature range of 4K-8K. At temperatures below 4K the hydrogen atoms spend such a large fraction of the time adsorbed on the walls that recombination, which takes place on the surface, rapidly depletes the sample. At temperature above 8K the vapor pressure of H_2 is often a problem. Between these limits is a wide operating range where H can be easily thermalized, forming what we shall describe as a helium temperature source. Among the immediate advantages of a helium temperature source compared to a nitrogen temperature or room temperature source are:

--Higher density for a given throughput
--Absence of molecular hydrogen in the beam
--Ease of cryopumping
--More efficient magnetic state selection.

Figure 1 shows a system which we have used for magnetic resonance studies of hydrogen in a H_2-coated container[2]. The aim was to create a flow of atomic hydrogen in a tube rather than to produce an atomic beam, but the design can be readily adapted to a beam source. An rf dissociator operates in liquid N_2, and the hydrogen emerges from a small aperture, ~1mm diameter, and passes to a helium-cooled region through a short thin-wall stainless steel tube. The hydrogen then flows to the resonance cell through a tube 9mm i.d. 17cm long, immersed in liquid He. Molecular hydrogen rapidly freezes to the surface. The measured flux to the bulb is 4 x 10^{17} atoms/sec; we believe that an appreciable fraction of the atoms recombine in the interconnecting tube, and that the flux near the source is higher. An impressive feature of this source is its simplicity. Employing no vacuum pump, it can run for hours solely by the cryopumping action of the walls. Eventually the inlet tube becomes thickly coated with frozen H_2 and must be warmed and pumped out.

Figure 2 shows an adaptation of the same design, miniaturized because of space constraints, which we are currently using at MIT as a source for creating spin-polarized hydrogen. The discharge operates at a density of about 10^{17} cm^{-3}, and the aperture is 0.7mm in diameter. The hydrogen flows through a 10cm long, 10mm diam., tube at a temperature of about 5K, and then passes through a 40cm long x 8 mm diam. tube to a cell at 0.3K. We cannot directly monitor the flux from the source, but the measured flow of (electron) spin-polarized atoms to the cell at 0.3K is 2 x 10^{15} atoms/sec. The actual flux from the source is probably much higher than this figure.

An alternative approach used by Walraven and Silvera[3] is to operate the discharge at room temperature and to let the atoms flow through a warm Teflon tube to a helium cooled thermalizer. They have obtained a flux from an accommodator at 8K of 2.4 x 10^{16} atoms/sec.

It should be possible to scale up these designs by a considerable factor without sacrificing their fundamental simplicity.

IIIB. Considerations in Designing a Helium Temperature Source

Analyzing the operation of an atomic hydrogen source from first principles is beyond the present state of the art, even for a conventional source. Nevertheless, one can estimate what changes might be expected in the governing factors when a source is adapted

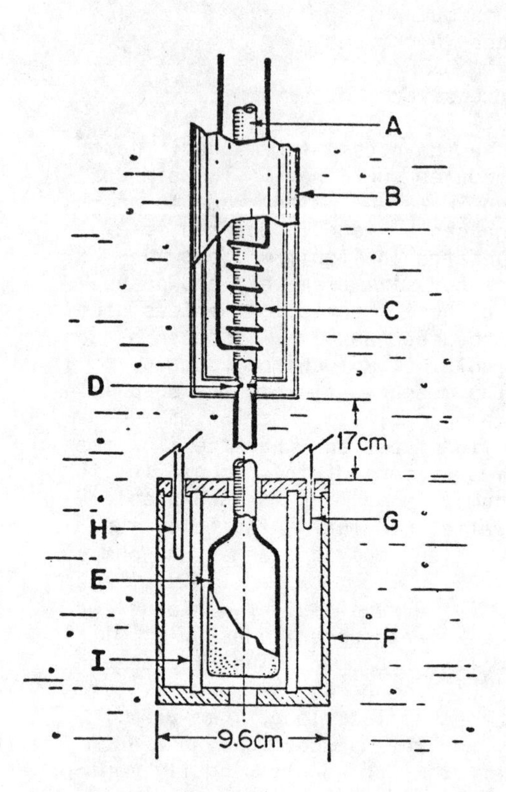

Fig. 1. Schematic diagram of apparatus for observing electron spin resonance on the hyperfine transition of hydrogen at liquid helium temperature. A, H_2 inlet B, liquid-nitrogen Dewar; C, dissociation coil; D, orifice; E, quartz storage bulb; F, microwave cavity; G, coupling loop; H, tuning rod; I, cylindrical capacitor. The apparatus is immersed in liquid helium.

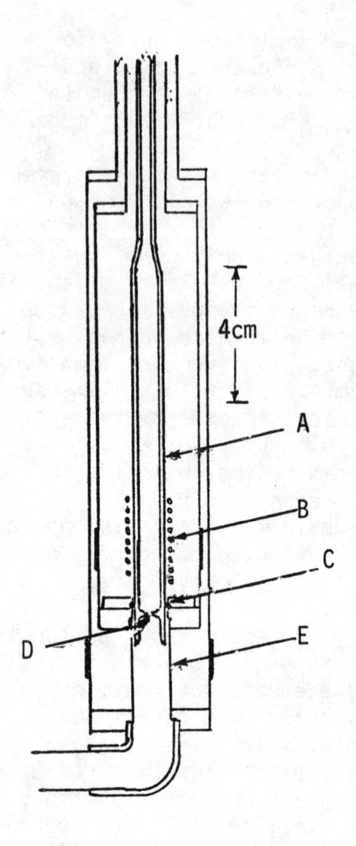

Fig. 2. Miniature dissociator used in the production of spin-polarized hydrogen. A, discharge tube, 9mm n.d. Pyrex; B, discharge coil, 8 turns copper; C, indium seal; D, orifice, 0.7mm diam.; E, 12mm diam x 0.25mm thick stainless tube. The inner chamber is filled with liquid N_2, and the apparatus is immersed in liquid helium. The dissociator dissipates about 3 watts of power at a frequency of 60MHz.

to helium temperature operation. For concreteness, we consider a hydrogen atomic beam source in which the electron is polarized by a hexapole magnet or similar state selector, and the nucleus is then polarized by a spin flip transition and second magnet. In a source for negative ions, the atom is finally converted to H^- by charge exchange. Such atomic beam sources are usually characterized by the flux of polarized atoms. A better figure of merit is the density of polarized atoms in the charge exchange region. The difference is significant when comparing room temperature and helium temperature sources.

Operation of a low temperature beam divides itself naturally into two topics: the source intensity, including flow and recombination rates, and the systematics of state selection by the magnetic deflectors. We discuss these in turn in the following two subsections.

1. Source Intensity

The useful intensity F of the (electron) polarized beam emerging from a hexapole state selector can be written

$$F = J(0)\Omega_{eff}f \tag{1}$$

where $J(\theta)$ is the flux per steradian from the source, Ω_{eff} is the effective acceptance angle of the state selector, and the factor f represents the fraction of atoms from the focusing magnet which are useful (i.e., have the correct nuclear spin, lie in the correct velocity range to be accepted by the ionizer, etc.).

In a conventional apparatus, the source is usually the dissociator itself. Assuming effusive flow through a thin aperture, the flux is

$$J_d(\theta) = \frac{A_d}{4\pi} n_d v_d \cos\theta \tag{2}$$

where n_d is the density, v_d is the mean speed, and A_d is the aperture area of the dissociator.

In a low temperature source, atoms from the dissociator pass into a short helium temperature thermalizer. If the fraction of atoms which "make it through" the thermalizer is ε_{th}, then the flux from the thermalizer is

$$F_{th} = \varepsilon_{th}F_d \tag{3}$$

where F_d is the total flux from the source. Losses in the thermalizer occur due to surface and volume recombination and escape through leakage areas. We can write $\varepsilon_{th} = \varepsilon_r \varepsilon_g$ where ε_r represents losses due to surface and volume recombination, and ε_g represents losses due to geometrical factors such as leakage. Although we cannot calculate ε_r accurately, it is possible to discuss the physical processes which limit it. The most critical of these is surface recombination, which is governed by the volume density.

The flux into the thermalizer is $A_d n_d v_d/4$. The flux out is $A_{th} n_{th} v_{th} \varepsilon/4$. Assuming that $\varepsilon \sim 1$, the mean density in the thermalizer is

$$n_{th} \simeq n_d \left(\frac{T_d}{T_{th}}\right)^{1/2} \frac{A_d}{A_{th}}. \tag{4}$$

The density and temperature of the thermalizer are governed by considerations of the recombination reaction $H+H+X \rightarrow H_2+X+4.6\text{eV}$. X stands for a third body or a surface. Morrow et al[4] have studied hydrogen recombination in helium-lined vessels. Their findings, which are discussed in Section IV B, are readily adapted to recombination of H or H_2. The principle result is that the hydrogen recombines primarily while adsorbed on the wall. When viewed in terms of the density of the gas, however, the hydrogen appears to decay by a two-body process according to

$$\dot{n} = -Kn^2. \tag{5}$$

The constant K is an effective two body rate defined below in Eq. (16); it includes a factor which takes into account the fraction of the time the atoms is adsorbed on the wall. The most important feature for our purposes is its dependence on the adsorption energy E_B:

$$K \sim \text{const}\ \exp(2E_B/kT). \tag{6}$$

The adsorption energy of hydrogen on H_2 has been measured by Crampton[5,6]: $E_B = 39.8(3)$K. This relatively high adsorption energy implies that K can be expected to vary rapidly with temperature in the helium temperature region.

The efficiency of the thermalizer depends on the competition between the recombination rate γ_r and the flow rate. From Eq. (5) we have $\gamma_r = Kn$. If we denote the flow rate constant by γ_f, then the probability that an atom will emerge from the thermalizer before it recombines is

$$\varepsilon_r = \gamma_f/(\gamma_f + \gamma_r). \tag{7}$$

γ_f is difficult to calculate if the system is in the intermediate region between hydrodynamic and molecular flow. If the flow is molecular, then

$$\gamma_f = A_{th}v_{th}/4V_{th} \text{, where } V_{th} \text{ is the volume in thermalizer, and} \tag{8}$$

where $v_{th} = \sqrt{8kT/\pi M}$ is the mean speed. By combining Eqs. 5-8, and using the results of IV B, the following expression can be obtained.

$$\varepsilon_r = [1 + \frac{8}{\sqrt{3}}\Lambda^2\lambda n N_t \exp(2E_B/kT)]^{-1}. \tag{9}$$

N_t is the mean number of adsorptive collisions an atom makes before escaping which may be considerably smaller than the actual number of wall collisions. Λ is the deBroglie wavelength, and λ is the surface collision range, as discussed in IV B. If we take N_t = 100, we obtain the results for ε_r shown in Table I.

T(K)	$n=10^{13}cm^{-3}$	$n=10^{14}cm^{-3}$	$n=10^{15}cm^{-3}$
4	.04	.004	
5	0.7	0.2	.025
6	0.97	0.8	0.3
7	0.99	0.95	0.65

Table 1. Transmission factor ε_r for a H_2 coated thermalizer, assuming 100 collisions.

These values have not been confirmed quantitatively, and they must be regarded with some skepticism. It is likely, however, that they err on the side of being too pessimistic. For instance, the source in reference 2 provides a significantly higher flux than Table 1 implies. This source involved flow with at least 400 wall collisions in a glass tube immersed in liquid helium. It is possible that the recombination energy slightly heated the H_2 surface, decreasing the recombination rate.

2. Magnetic State Selection, Ω_{eff}

The collection efficiency ~~of~~ a conventional magnetic state selector is proportional to the ratio of magnetic to kinetic energy. For a hexapole state selector with a small input angle

$$\Omega_{eff} \simeq 2.1\frac{\mu_o B}{kT} \qquad (10)$$

where μ_o is the Bohr magneton and B is the field at the pole tip. For B = 10 kG and T = 300K, $\Omega_{eff} \simeq 4\times10^{-3}$. Because Ω varies inversely with temperature, one would naively expect an improvement by a factor of 60 if a room temperature source is converted to 5K operation. However, to make use of such a large acceptance angle the thermalizer aperture would have to very close to the magnet, if not inside. Steps would have to be taken to maintain the pumping speed of the system, either by opening the magnetic structure or perhaps cryopumping on the magnet. Since scattering near the source can ultimately limit the throughput, the temperature dependence of the total cross section must also be considered. Fortunately, recent calculations by Morrow and Berlinsky[7], Fig. 3, reveal that the cross section is smaller at 5K than at 80K or 300K. These results contradict earlier calculations of Allison and Smith[8], which use an incorrect formula[9].

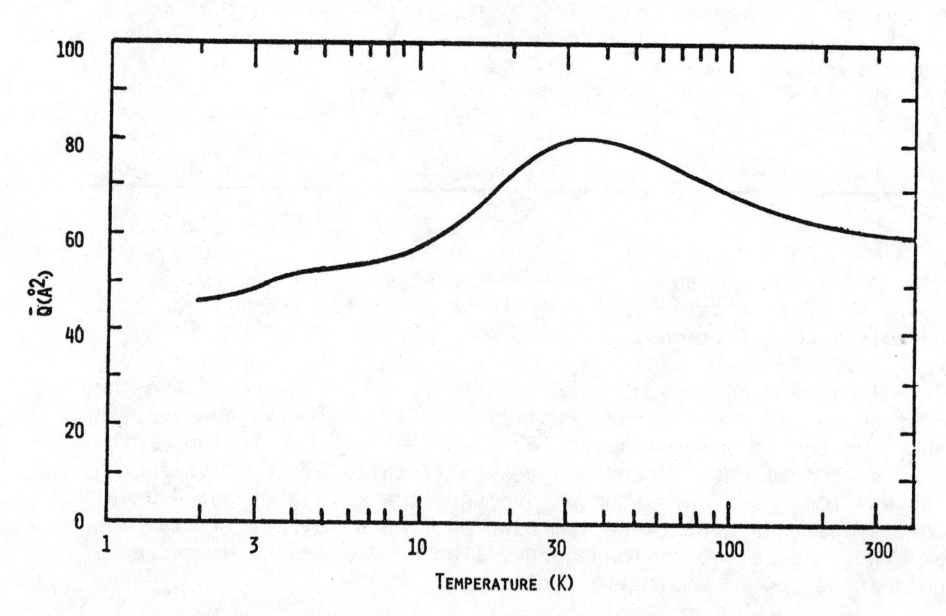

Fig. 3. Thermally averaged total scattering cross section for hydrogen. (Courtesy of M. Morrow and A.J. Berlinsky.)

Because a low temperature source requires much weaker deflecting fields than a conventional source, there is considerable latitude in configuring the state selecting magnets. The improvements to be expected if a source is adapted for low temperature operation depend strongly on the configuration. Consider first an ideal system in which the flux is limited only by scattering at the source and for which the state selectors are optimally designed. If this system is converted to low temperature operation simply by reducing the magnetic field in proportion to the temperature, no appreciable gain is to be expected. More precisely, the beam density will scale inversely with the scattering cross section, producing a very modest gain in density at low temperature.

A more realistic case is one in which the total flux is limited by the pumping speed of the system. In such a case large improvements are to be expected. For a fixed throughput the beam density increases as $1/\sqrt{T}$. Further, the pumping speed itself may be substantially increased by the use of cryopumping. The design of magnetic state selectors for a low temperature polarized proton ion source has not been analyzed in detail so far, but one may expect appreciably better performance than for high temperature beams.

Crampton[6] has studied the design criteria for a low temperature source for a hydrogen maser. He considered atoms at a temperature of 7K focused to a circular aperture 20cm downstream from the magnet with the same diameter as the magnet gap, 1.3cm. He calculated a gain in Ω_{eff} of 15 compared to the room temperature source. The gain in the beam density is only 7.5, however, due to the larger area of the low temperature beam. Although such a gain is modest, it is by no means negligible. It seems probable that one can achieve even higher gains with improved design.

In summary, the helium temperature regime provides a number of attractions for conventional hydrogen atomic beam sources: higher density, more efficient state selection, and the advantages of cryopumping. It is difficult to estimate the extent of the improvements for a particular system short of going through a detailed analysis, and certainly such an analysis is warranted if renovations are planned for an existing apparatus, or new apparatus is being considered. In the latter case, however, it is imperative to consider the opportunities provided by a low temperature polarized-proton source, which is the subject we turn to next.

IV. VERY COLD REGION, T ∽ 0.3K

IV A. General Considerations

At very low temperatures it is possible to achieve conditions in which the magnetic spin energy of hydrogen in an applied field is large compared to the thermal energy, reversing the usual state of affairs. Under such conditions novel hydrogen polarization techniques are possible. 100% electron polarization has been achieved in a simple flow system, with density greater than 10^{17} cm^{-3}. Furthermore, in a recent development it has been found that under relatively simple experimental conditions electron-spin-polarized hydrogen can acquire a high nuclear polarization. Conditions in which the nuclear polarization is greater than 99% have been achieved. Many applications are possible: the method can provide a medium of unpredecented density for producing proton-polarized H^-; it can provide a dense proton-polarized atomic beam for use as a jet or as a collision medium, and the gas may be useful for a polarized proton target. The proton-polarized hydrogen can be compressed to much higher density in a pulsed mode.

The energy of an electron spin in a magnetic field B, expressed as a temperature, is $\pm\mu_o B/k = \pm 0.67B\,^\circ K$ where B is in tesla. The equilibrium electron polarization is

$$P_e = \frac{n_+ - n_-}{n_+ + n_-} \simeq \exp(-2\mu_o B/kT) - 1. \tag{11}$$

where + and - refer to the states $m_s = +1/2$ and $-1/2$, respectively. In a 10T field at 0.3K the "spin impurity", $n_+/n_- \simeq \exp(-2\mu_o B/kT)$, is 3×10^{-20}; the system is 100% electron-polarized.

In conventional polarized proton ion sources, creating a high electron polarization in hydrogen is merely an intermediate step toward achieving high proton polarization. From the point of view of high density cold hydrogen technology, however, achieving a high electron polarization is vital to assure that atomic collisions occur only in the molecular triplet state. (The triplet potential is essentially repulsive, and the collisions do not lead to molecular recombination.) The need to prohibit recombination puts stringent requirement on the degree of electron polarization. Fortunately, as the figures quoted above suggest, at low temperature and high magnetic field extremely high polarization is achieveable. It is for this reason that all schemes which have so far attained high densities of atomic hydrogen have these two elements in common.

A third common element to methods for producing cold hydrogen is the use of a helium film to inhibit recombination on the surfaces of the container. Ideally one would trap the atoms by magnetic fields alone without the use of any confining surfaces. Such a trap is not possible, however, at least not with a static magnetic field, since it would require a field maximum in free space. Dynamical traps may be possible, though no practical designs have so far been proposed. Thus confining surfaces are needed; helium is the only serious candidate. To arrive at this conclusion one need only consider the adsorption energy for hydrogen on the surface. For a ^{4}He surface it is 1K, for ^{3}He it is 0.44K. The next best candidates is H_2, for which the adsorption energy is 39K. With such a strong surface attraction the atoms would rapidly adsorb and recombine. From the point of view of adsorption energy, ^{3}He is preferable to ^{4}He. However, the latter species has the important practical advantage of being superfluid at 0.3K, creating a perfect surface coating in which wounds are immediately healed.

An apparatus which illustrates a method for achieving high density electron and proton spin-polarized hydrogen is shown in Fig. 4. The dissociator shown in Fig. 2 is connected to the inlet at the top.

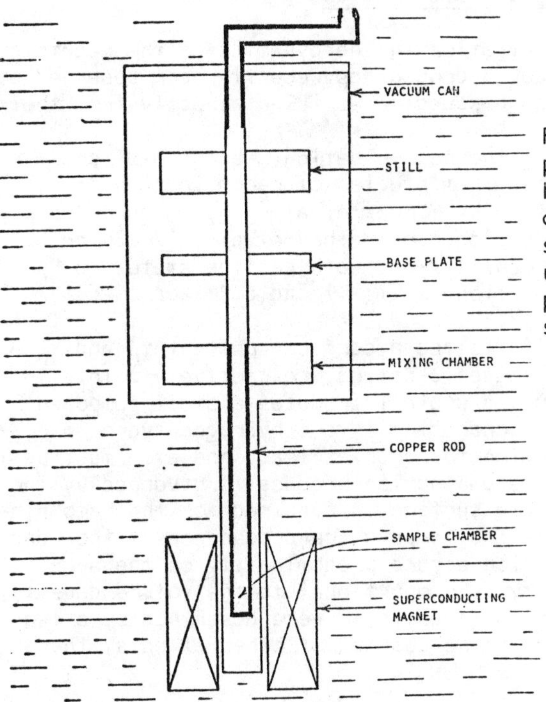

Fig. 4. Apparatus for producing spin-polarized hydrogen. The pressure of atomic hydrogen in the sample chamber is measured with a capacitive pressure transducer (not shown).

The sample chamber is located at the center of a superconducting solenoid which can provide a field up to 11T. The bottom of the cell contains a pool of ^{4}He. A saturated superfluid film completely covers the cell and flows up the inlet tube to the still of the refrigerator where it is thermalized in a heat exchanger at 1K. The helium continues to flow toward the 4K region, and eventually boils and refluxes. This arrangement places most of the heat load of the superfluid film flow on the still rather than on the mixing chamber.

The incoming hydrogen is thermalized at 0.3K before reaching the magnet. Atoms in the spin "up" states are repulsed by the 7K magnetic potential barrier; atoms in the spin "down" state are attracted into the magnetic potential well. As they enter they are continually thermalized by wall collisions. Once in the well, the escape rate is so low that the atoms are, to first approximation, permanently trapped. The density increases until it is finally limited by recombination processes. It is this recombination process which creates the proton polarization.

In section IV B, we discuss the dynamics of recombination and relaxation which govern the creation of the proton-polarized gas, and in section IV C, we describe some possible applications.

IV B. Recombination and Proton Polarization

The role of proton polarization in the dynamics of the recombination process of spin polarized hydrogen has recently been observed by Cline et al[10]. The following discussion is taken directly from their work.

A clear understanding of the recombination mechanism of spin-polarized hydrogen has emerged from studies of recombination of H on He films at zero magnetic field by Morrow et al[(4)] and measurements of the decay of spin-polarized hydrogen by Mathey et al[(11)]. Spin-polarized hydrogen normally consists of two hyperfine states, a "pure" state $|-1/2,-1/2\rangle$ (in the notation $|m_e,m_p\rangle$), and a "mixed" state $\cos\theta|-1/2,1/2\rangle - \sin\theta|1/2,-1/2\rangle$, where $\tan 2\theta(B) = a/[h(\gamma_e + \gamma_p)B]$. (a is the hyperfine constant, γ_e and γ_p are the electron and proton gyromagnetic ratios, respectively.) In a 10T field, $\sin\theta \approx 2\times10^{-3}$. The mixed state introduces a small component ($\approx\sin\theta$) of electronic singlet character into a hydrogen-hydrogen collision. Thus, for recombination to occur at least one atom must be in the mixed state so that the recombination process is governed by the proton spin. A third body or a surface is required for the recombination reaction, and studies of the temperature dependence of the reaction rate show that the reaction occurs predominantly on the surface[4,11]. The surface density is proportional to the volume density, however, and the reaction proceeds as if it were described by a two-body gas collision process. In the absence of other effects, the mixed state would disappear by molecular reaction, leaving the gas in the pure state with 100% proton polarization.

We must also consider the effect of nuclear relaxation. The rate T_1^{-1} is proportional to the total density n_p+n_m and causes the population difference n_p-n_m to decay to zero. (n_p and n_m are the pure and mixed state densities, respectively.) The time evolution of the densities is governed by the following equations[12]:

$$\dot{n}_p = F_p/V - \gamma_o n_p - K n_p n_m - G(n_p+n_m)(n_p-n_m) \tag{12a}$$

$$\dot{n}_m = F_m/V - \gamma_o n_m - K n_m(n_p+2n_m) + G(n_p+n_m)(n_p-n_m) \tag{12b}$$

V is the effective confinement volume and F_p and F_m are the fluxes of pure and mixed state into the cell, respectively. We assume that $F_p = F_m$ since the two states are created in equal numbers at the source and the nuclear spin relaxation time is long compared to the transit time from the source to the cell. γ_o is the rate of one body decay processes such as escape from the magnetic potential well. K represents the two-body recombination process. (A term representing three-body recombination in the gas phase has been omitted since we find no evidence for such a process at our densities.) $G(n_p+n_m) \equiv T_1^{-1}$ is the nuclear spin relaxation rate which can have contributions from transitions taking place in the gas phase[12,13] or on the surface[14].

Assume that $K \gg G$ and that the cell can be filled in a time short compared to $(Kn_+)^{-1}$. Initially $n_p \simeq n_m$ and the total density begins to decay rapidly according to the equation

$$\dot{n}_+ \simeq -\gamma_o n_+ - K n_+^2. \tag{13}$$

After a time long compared to $(Kn_+)^{-1}$, most of the mixed state has disappeared and the decay of n_+ becomes much slower since it is limited by the rate at which the pure state is converted to the mixed state:

$$\dot{n}_+ \simeq -\gamma_o n_+ - 2G n_+^2. \tag{14}$$

In this limit the remaining gas has a high nuclear polarization

$$P_n = \frac{n_p - n_m}{n_+} \cong 1 - 2\frac{G}{K}. \tag{15}$$

An experimental plot of the time evolution of the density of spin polarized hydrogen, Fig. 5, illustrates this description. (The density is determined by measuring the gas pressure.) The caption explains the sequence of events. Fig. 6 shows experimental values for the effective recombination constant K and the nuclear relaxation constant G.

In equilibrium after long times the nuclear polarization is given by Eq. 15. At 300mK the polarization is calculated to be 98%; at 240mK it is 99.4%. These results are particularly striking in view of the unconventional nature of the polarization mechanism: chemical reaction.

Behavior of the rate constants

In order to apply these results to other configurations it is necessary to describe the dependence of the recombination constant K and nuclear relaxation constant G on the experimental parameters. The effective two body rate constant K is related to the intrinsic surface rate constant K_s' by the expression

$$K = K_s(A/V)\Lambda^2 \exp[2E_B/kT] \qquad (16)$$

where A and V are the area and volume of the confinement region, $\Lambda = (h^2/2\pi MkT)^{1/2}$ is the thermal deBroglie wavelength and E_B is the binding energy of the atomic hydrogen on the liquid helium surface. $K_s \equiv \lambda v_s$, where the length λ is the analog of a collision cross section for a two dimensional system and $v_s = (32kT/3\pi M)^{1/2}$ is the mean thermal speed on the surface[4]. λ is not expected to have a strong temperature dependence. Assuming that λ is constant so that the slope of $\ln(KT^{1/2})$ varies as 1/T, then we find from the data shown in Fig. 6 that $E_B = 1.01(6)$K.

From Eq. (16) we can obtain a value for the intrinsic surface recombination rate constant K_s. Using our value of $A/V = 4.9\text{cm}^{-1}$, we find $K_sT^{-1/2} = 4.5(3)\times10^{-10}\ \text{cm}^2\text{s}^{-1}\text{K}^{-1/2}$ at a field of 11T, which gives

$$K_sT^{-1/2}B^2 = 5.4\times10^{-8}\text{cm}^2\text{s}^{-1}\text{K}^{-1/2}\text{T}^2$$

Turning now to the influence of nuclear polarization on the decay, the temperature dependence of G (see Fig. 6) reveals that gas and surface phase relaxations are both important. The rate constant which appears in Eq. 12 is thus a sum of two terms:

$$G = G_g + G_{eff}. \qquad (17)$$

Fig. 5. Trace of experimental decay curve taken at 300mK and 11T. The noise is less than the width of the line. A) The hydrogen source is turned on. B) The source is turned off and the density decays rapidly as the mixed state recombines. C) The system is now left in the pure state. It decays slowly due to nuclear relaxation to the mixed state. D) The sample is destroyed by recombining the hydrogen on a bolometer. E) The system is reloaded to the same density as C. Note that the decay rate is much greater than at C due to the presence of the mixed state. (from ref. 10)

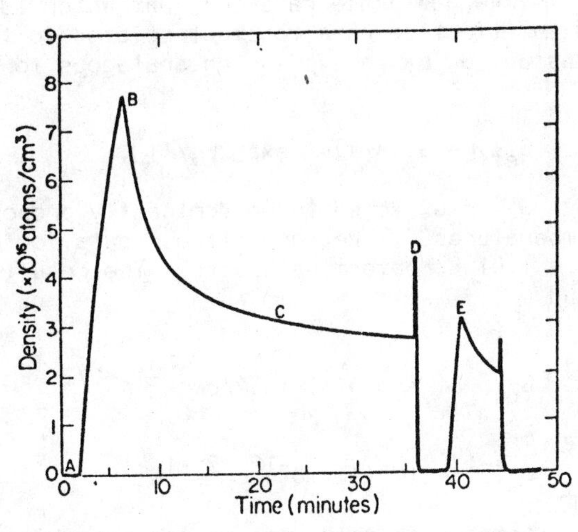

Fig. 6. Measurements of the recombination coefficient K and the nuclear relaxation coefficient G. Logarithmic plot of $KT^{1/2}$ (circles, left axis) and linear plot of $GT^{-1/2}$ (triangles, right axis) vs. 1/T. Measurements were made at a field of 11T. The solid lines are least squares fit to the data. The dashed line indicates the gas phase contribution to the nuclear relaxation. (from ref.10)

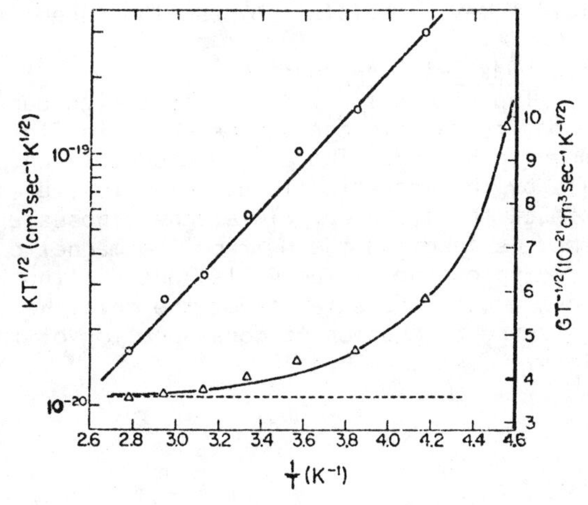

G_g is the gas phase rate constant which is proportional to $T^{1/2}$. G_{eff} is an effective rate constant related to the intrinsic surface rate constant G_s by an expression analagous to Eq. (16):

$$G_{eff} = G_s(A/V)\Lambda^2 \exp[2E_B/kT]. \tag{18}$$

G_s is expected to be dominantly proportional to T in our range of temperatures[14]. We have fit our data for G using Eq. 18 and the value $E_B = 1.01$ K, determined above. The results are plotted in Fig. 6. We find

$$G_g T^{-1/2} = 3.5(3)\times10^{-21}\text{cm}^3\text{s}^{-1}\text{K}^{-1/2} \tag{19}$$

$$G_s T^{-1} = 1.7(4)\times10^{-12}\text{cm}^2\text{s}^{-1}\text{K}^{-1}. \tag{20}$$

Armed with this data on the recombination and relaxation rates, we turn now to some possible applications of the proton-polarized gas.

IV.C Application to a Polarized Proton Ion Source

The ability to produce high densities of hydrogen with polarized protons offers numerous new opportunities for producing polarized proton ion sources, targets and jets. To suggest some of the possibilities it may be helpful to describe three lines of approach.

C1. "Gas Cell" Approach

The simplest way to obtain a high density of polarized atoms is to modify the storage scheme shown in Fig. 4. A schematic sketch is shown in Fig. 7. The gas is stored in an open cylinder, confined axially by the magnetic field. Fast ions used for charge exchange pass through the cylinder. This scheme poses a number of technical problems: extracting the H^- from the magnetic field, avoiding charge build-up on the helium walls, and dealing with the helium film which flows around the exterior of the cell. We shall not address these, but limit the discussion to consideration of the basic operating conditions.

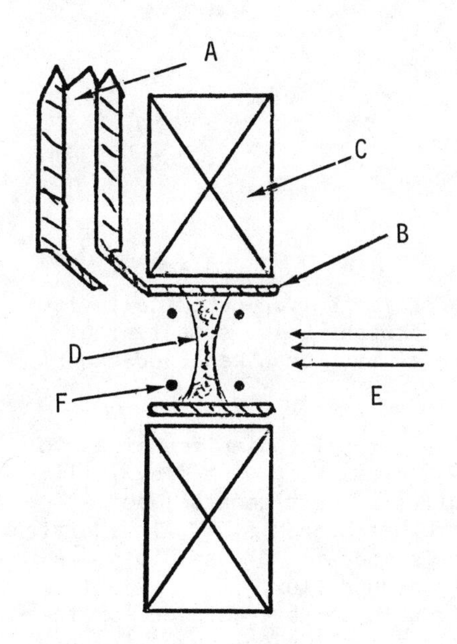

Fig. 7. Schematic diagram of polarized proton ion source. A, hydrogen inlet, from refrigerator; B, annular storage cell; C, superconducting solenoid; D, gas of magnetically confined proton-polarized hydrogen; E, fast Cs beam; F, extraction electrodes for H^- ions. Sections A and B are coated with a helium film at 0.35K; the extraction electrodes can be at higher temperature. The magnetic field is shaped to give a sharp maximum along the axis. Because of the high atomic density, the collision and extraction region is short compared to a conventional source.

In the "gas cell" approach a stationary flow situation is achieved in which the densities of the pure and mixed states, n_p and n_m respectively, are constant. Closed solutions to Eqs. 12 are possible if the single body decay rate, γ_o, can be neglected compared to the other decay terms, which is often an excellent approximation. In this case the following results are obtained for the total density n_+ and the nuclear polarization P_p:

$$n_+ = \left(\frac{F}{2KVy}\right)^{1/2} \tag{21}$$

$$P_p = 1 - y. \tag{22}$$

Here F is the total flux of (electron) spin-polarized atoms, K is the effective two-body decay constant, and V is the volume of the cell. The parameter y is given by

$$y = -\beta + \sqrt{\beta}\sqrt{1+\beta}. \tag{23}$$

where $\beta = G/K$. In the limit G<<K, which is usually a good approximation, we have

$$n_+ \simeq \left(\frac{F}{2KV}\right)^{1/2}\left(\frac{K}{G}\right)^{1/4} \quad (24)$$

$$P_p \simeq 1-2\left(\frac{G}{K}\right)^{1/2} \quad (25)$$

More generally, Eqs. 12a and 12b can be solved numerically for any desired values of the constant. We present below some typical results based on conditions which have already been realized experimentally. These are:

Flux: Our source gives a total flux of spin-polarized atoms to the cell of approximately 2×10^{15} atoms s^{-1}. The flux is limited by thermal loading of the dilution refrigerator; better thermal design should allow a somewhat higher flux. In addition, the capacity of the refrigerator is modest: 3mW at 0.3 K. A larger refrigerator would allow a higher flux.

The ultimate limit to the flux is set by the need for the refrigerator to remove all the recombination energy of the atoms (assuming that the hydrogen is all fated to recombine in the cell). The recombination energy, 2.4eV per atom, gives a limiting flux of 2.6×10^{15} atoms s^{-1} for each milliwatt of refrigeration power.

The flow equations involve the flux per unit volume of the storage cell. In our present design, a cylinder 1 cm in diameter, the effective volume is 5 cm^3.

Single Body Relaxation Rate, γ_o: Atoms escape from the cell by molecular effusion, retarded by the magnetic potential barrier. For an open cylindrical geometry it is readily shown that

$$\gamma_o = \frac{1}{4}\frac{\bar{v}}{L} e^{-\mu_o B/kT} \quad (26)$$

where $\bar{v} = \sqrt{4kT/\pi M}$ is the mean thermal speed, and L is an effective length of the cell given by

$$L = \int \exp[-\mu_o B(z)/kT] dz \quad (27)$$

The integration along the axis extends between end surfaces, or to $\pm\infty$ in the case of an open cell. Although L depends on the magnetic field, it is a reasonable approximation to treat it as a constant. We have taken $L = 5cm$.

In practice, γ_0 is extremely small except at low magnetic field, 2T or less.

Recombination Coefficient K: The dependence of K on the magnetic field and temperature is described by Eqs. 16. In addition, it depends on the geometry through the surface-to-volume ratio, A/V. In our cell design there is an appreciable amount of surface due to the pressure transducer, and $A/V = 5cm^{-1}$. For an open cylindrical cell, $A/V = 2/R$.

Relaxation Constant G: At higher temperatures, 0.3K or above, G is dominated by the volume contribution, Eq. 19. At lower temperatures the surface contribution, Eq. 20, is important. The latter depends on geometry in the same way as K.

Results

Values for the density and proton polarization as a function of magnetic fields are shown in Figs. 8a and 8b, respectively. A striking feature is the relatively slow variation of density with field; a factor of approximately two between 5T and 11T. Escape from the magnetic well becomes important below 2T, causing the polarization to fall abruptly. The slight fall in polarization at high field is due to the decrease with field of the ratio G/K. There is a trade-off between polarization and density; the highest temperature, 0.35K, gives the largest density and smallest polarization. Nevertheless, even at 0.35K the polarization is over 83% at 11T, and over 86% at 5T.

As a general rule, the density scales approximately as the square root of the flux-to-volume ratio.

For certain applications the densities in Fig. (8) are too high by many orders of magnitude. One way to decrease the density is to have a field gradient across the cell so that the polarization is established in a high field region, while the atoms are employed for collisions or other purposes in a lower field, lower density region.

C2. Polarized Atomic Beam Source

The techniques described above can be modified to serve as a polarized atomic beam source. Because the atoms are in a pure spin state, both the electron and proton polarization is preserved when the atoms flow to a low field region. The flux can be much higher than in conventional atomic beam devices.

The basic idea is to use the "gas cell" approach described above to polarize the hydrogen and to let the gas escape from the trap at some preselected rate. In effect, the gas cell serves as a source for

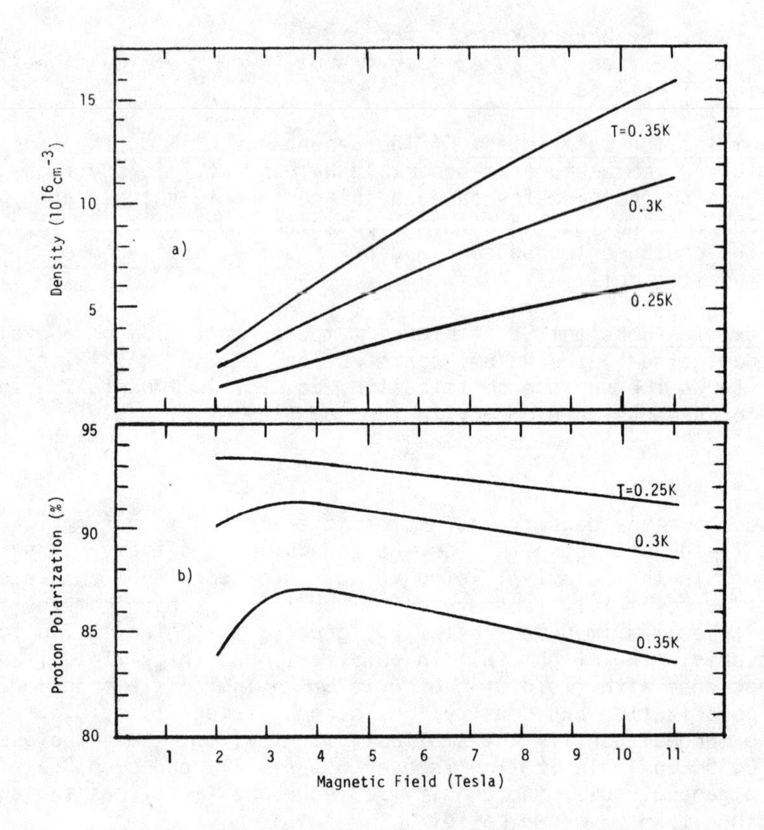

Fig. 8. Equilibrium density (a) and proton polarization (b) of hydrogen in a gas cell for the temperatures indicated. Incident flux, 1.5×10^{15} atoms s^{-1}; volume of cell, 5 cm^{3}; volume-to-area ratio, 0.2 cm.

the beam, a source of fully polarized atoms.

Ironically, the principle experimental challenge is not to create a high density but to extract the atoms from the cell. The simplest approach, lowering the magnetic field, is not satisfactory since the field might have to be so low that the proton polarization would be seriously degraded. Various extraction schemes are possible such as locally heating atoms near the edge of the magnet. Without specifying the extraction mechanism we shall simply assign it a rate constant γ_e, so that the total one-body decay rate is $\gamma_o + \gamma_e$. The flux out is given by

$$F_{out} = \gamma_e nV. \tag{28}$$

Because the flux and the polarization depend on several parameters we shall simply make a few general comments rather than attempt to make a systematic survey of operating conditions.

There is usually a tradeoff between output flux and the polarization. In many cases the ratio F_{out}/F_{in} can slightly exceed 20% with a proton polarization of 80%. For $F_{out}/F_{in} \simeq 10\%$, P_p increases to about 85%. If only a few percent of the incident flux is extracted, P_p can reach 90%. These results hold for a field of 4T. Slightly better operation can be expected at lower field, though a "preselector" to exclude atoms in the upper hyperfine states might be required. An operating temperature of 0.35K is adequate; a somewhat lower temperature can be used to achieve higher polarization at the expense of some flux. These results are independent of the input flux over a wide range, though the optimal value of γ_e depends on the flux. Typically, γ_e should be in the range of 10^{-2} to $10^{-3}\ s^{-1}$. As one example, with an input flux of $3 \times 10^{15}\ s^{-1}$, a flux of $7 \times 10^{14}\ s^{-1}$ can be extracted with a proton polarization of 79%, and $3 \times 10^{14}\ s^{-1}$ can be extracted with a polarization of 87%.

With a 100 milliwatt refrigerator an input flux greater than $10^{17}\ s^{-1}$ should be obtainable. A flux greater than $2 \times 10^{16}\ s^{-1}$ could be extracted with 80% polarization. If the atoms emerge from a 3mm diam aperture, the density at the aperture is approximately $3 \times 10^{13}\ cm^{-3}$.

C3. Pulsed Target

The density of polarized atoms in a cell is limited by two-body processes, relaxation and recombination. The equilibrium density varies with the square root of the flux while the load on the refrigerator varies linearly with the flux. Thus it is difficult to achieve much higher equilibrium density by a "brute force" increase of beam intensity. In a pulsed mode, however, very high density should be possible. Once the gas in a cell has achieved a high proton polarization due to recombination of the mixed state as in Fig. 6, the remaining atoms can be compressed to high density. Figure 10 illustrates a possible design. The spin relaxation rate increases with density, causing the atoms to recombine faster, but there should be adequate time for many applications before the gas is consumed. From Eq. (14), and neglecting γ_O, it can be seen that the characteristic decay time is

$$\tau = (2Gn)^{-1} \qquad (29)$$

Using the value $G = 2.2 \times 10^{-21}\ cm^3 s^{-1}$, we obtain

$$\tau = 2.2 \times 10^{20} n^{-1} s^{-1}. \qquad (30)$$

This indicates that a lifetime of 2 seconds should be achieveable at a density of $10^{20}\ cm^{-3}$. It should be possible to compress the sample in a small fraction of a second, and densities greater than $10^{20}\ cm^{-3}$ may be achieveable. Of course, such large extrapolations must be viewed as somewhat tentative. Experiments on compressing a nuclear-spin-polarized gas of atomic hydrogen are underway at M.I.T.

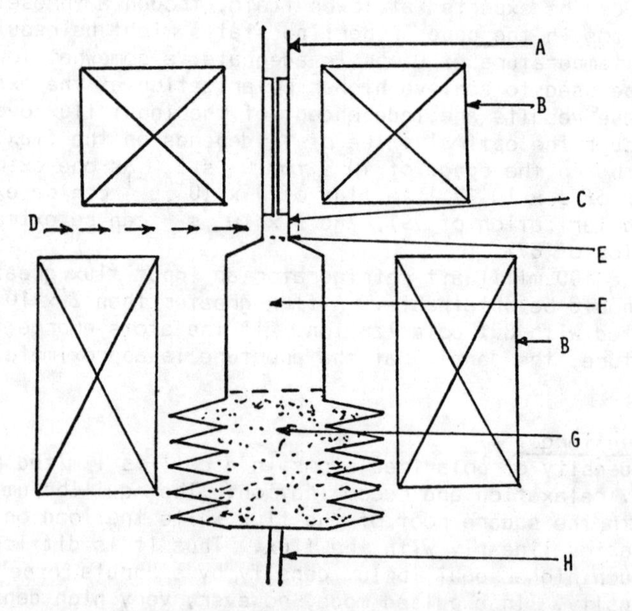

Fig. 10. Proposed atomic hydrogen polarized proton target. A, hydrogen inlet (from dilution refrigerator); B, superconducting solenoid; C, thin wall scattering chamber; D, high energy proton beam; E, level of liquid helium reservoir in compressed position; F, storage cell; G, liquid helium reservoir; H, shaft to compression drive mechanism.

ACKNOWLEDGEMENTS

Much of the work described here is the result of a collaborative effort with my colleagues, R.W. Cline and T.J. Greytak. I would like to thank them for their comments on the manuscript. I would also like to thank S.B. Crampton for helpful discussions on the source design, and A.J. Berlinsky for recalculating the hydrogen kinetic cross section. The research carried out at MIT was supported by the National Science Foundation, Grant DMR 80-07850.

REFERENCES

1. L.H. Nosanow, Jour. Physique, 41, (Colloq. C7), C7-1, (1980).
2. S.B. Crampton, T.J. Greytak, D. Kleppner, W.D. Phillips, D.A. Smith, and A. Weinrib, Phys. Rev. Lett. 42, 1039, (1979).
3. J.T.M. Walraven and I.F. Silvera, Rev. Sci. Inst., to be published.
4. M. Morrow, R. Jochemsen, A.J. Berlinsky, and W.N. Hardy, Phys. Rev. Lett. 46, 195 (1981) and 47, 455 (1981).
5. S.B. Crampton, Jour. Physique, 41, (Colloq C7), C7-249 (1981).
6. S.B. Crampton, J.J. Krapczak and S.P. Souza, Conference on Metrology, Aussois, October, 1981 (Jour. Physique, to be published).
7. M. Morrow and A.J. Berlinsky, private communication.
8. A.C. Allison and F.J. Smith, Atomic Data 3, 317, (1971).
9. A.J. Berlinsky, private communicaton.
10. R.W. Cline, T.J. Greytak and D. Kleppner, Phys. Rev. Lett. 47, 1195 (1981).
11. A.P.M. Matthey, J.T.M. Walraven, and I.F. Silvera, Phys. Rev. Lett. 46, 668 (1981).
12. B.W. Statt and A.J. Berlinsky, Phys. Rev. Lett. 45, 2105 (1980).
13. E.D. Siggia and A.E. Ruckenstein, Phys. Rev. B 23, 3580 (1981).
14. E.D. Siggia and A.E. Ruckenstein, private communication.

Report on the Cold Atomic Hydrogen PPIS Work Session I

Thursday May 21, 1981

Chairman: D. Kleppner Secretary: L. Levy

The question addressed in this session is to what extent the intensity of polarized proton sources could be improved by using a low temperature hydrogen source at the atomic stage, as discussed by D. Kleppner.

Presently the existing polarized sources deliver about 100 μA of positive polarized beam. A goal of two orders of magnitude improvement in the intensity would be desirable, though more would not be useful in the present accelerators. The workshop decided to limit the discussion to improvements to a conventional source, leaving novel approaches to the next session.

A conventional approach at lower temperature would be of the type

$$\begin{matrix}\text{cold}\\ \text{source}\end{matrix} \rightarrow I_s \rightarrow \begin{matrix}\text{state}\\ \text{selector}\end{matrix} \rightarrow \vec{I}_p \text{ , where}$$

I_s = atomic beam intensity

$\vec{I}_p$ = polarized beam intensity.

The polarized beam intensity $\vec{I}_p$ is related to I_s by

$$\vec{I}_p = \frac{I_s\,\Omega_{eff}}{\pi}\, f \propto n_p\, \bar{v}_p\, A_p. \tag{1}$$

Ω_{eff} is the effective solid angle of acceptance of the focussing magnet which in the limit of small angles is given by

$$\Omega_{eff} = 2.1\,\frac{\mu B}{kT} \tag{2}$$

and f is a focussing factor: it represents the fraction of atoms coming out of the focussing magnet which can be used, and in particular includes the factor of 1/2 due to the state selector.

From equation (2) one would believe that an improvement to be obtained by using a low temperature source would be to increase the effective acceptance angle by the ratio of the source temperature to room temperature:

$$\frac{300K}{4°K} = 75$$

leading to an acceptance angle of ~ 22.7°. This angle being so large one might wonder if one can realistically obtain such an improvement. In particular it would imply that the source would be right at the entrance of the magnet. Another important remark was made indicating that because of the larger angular spread at the source, the brightness of the source after focussing would not be increased. In that respect the hexapole focussing might not be the best.

The question of the intrinsic atomic beam intensity was then addressed: typical intensities for room temperature sources are 10^{19}/sec [1 torr-liter/sec]. Because the scattering becomes more severe at lower temperatures, the gain in density n_p of $T^{-1/2}$ due to the velocity $\bar{v}_p$ is lost on the intensity

$$I_s = \bar{v}_s \cdot \frac{n_s A}{4} .$$

As scattering limits the density n_s to a constant, I_s would be decreased by $T^{1/2}$, leaving n_s no net gain in the $I_s/\bar{v}_s$ factor.

The various factors discussed are summarized in Table 1.

<u>Worksession Conclusions</u>

1. A gain of at least

$$\left(\frac{T_{room\ temp}}{T_{4°K}}\right)^{1/2} = 8.7$$

is to be expected from a low temperature source, and possibly more if focusing optics are improved. Indeed, because of the larger acceptance angle, the optics improvements become more important at low temperature. In particular a solenoid focusing may have large advantages as:

1. scattering would be reduced

2. a better geometrical acceptance [f] would be possible

3. the largest possible acceptance angle would be obtained, provided that fringing defocusing would be small.

The scattering at the source limits the beam flux and leads to no gain at lower temperature I_s/v_s.

Conclusion

Gain of factor $(T[room]/T[He])^{1/2}$ = 8.7 seems quite realistic. A somewhat larger gain may be possible using the conventional approach. A much larger gain may be possible with a non-conventional source to be discussed at the next workshop.

TABLE 1: Gain Factor Estimates for 4°K Source Compared to Room Temperature Source

Room Temperature Source	4°K	Gain Factor in n_p	Comment
$\Omega_{eff}=3\times10^{-3}$ sr	.22	75 (max)	-source must be at the magnet
		8.7 (min)	-larger angular spread reduces brightness
$\overline{v}_p$ $=3\times10^5$cm/sec	3.7×10^4cm/sec	8.5	-does not constitute a real gain because of the reduction of I_s
I_s $=10^{19}$/sec	$\frac{v_s n_s A}{4}$	$\frac{1}{8.5}$	-scattering limits the density to a constant, so the throughput is lower
$n_p \propto \frac{I_s \Omega_{eff} f}{\pi \bar{v}_p A_p}$		8.7 (min)	

Report on the Cold Atomic Hydrogen PPIS Work Session II

Friday, May 22, 1981

Chairman: D. Kleppner   Secretary: L. Levy

The topics discussed in this section are the non-conventional source proposed by D. Kleppner with a large solenoid field, its use for spin polarized hydrogen and the possibilities of storing polarized hydrogen at the ionization stage.

1. Solenoid Low Temperature Source

The source proposed can be schematically shown as follows:

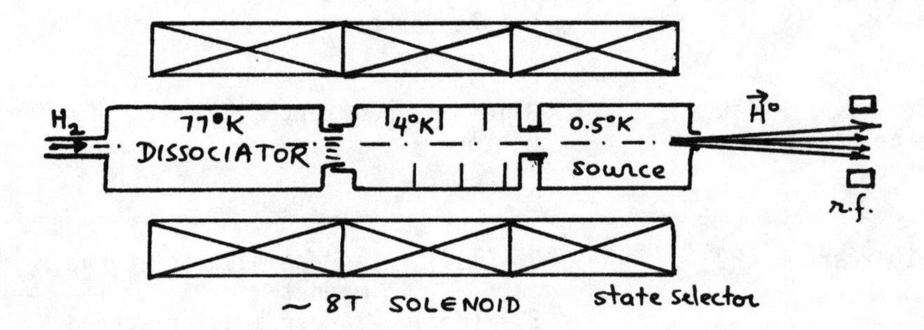

The polarized intensity can be written, as discussed in the previous session as

$$I_p = \frac{I_s \Omega f}{\pi}$$

and the density

$$n_p = \frac{I_s \Omega f}{\pi \bar{v}_p A_p} .$$

Because the atoms leaving the magnetic field of 100 kG are accelerated the emission of the source is not of a traditional effusion type but becomes substantially narrower.

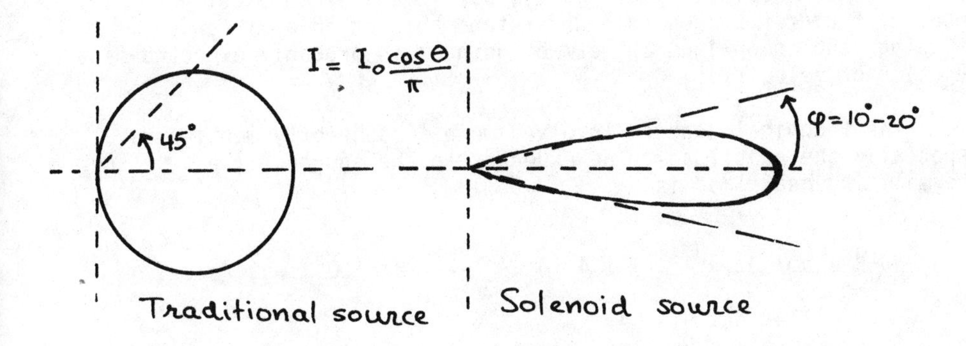

The 8°K longitudinal acceleration produces a beam of 10 to 20° divergence.

Another important difference comes from the velocity distribution which is shifted up in velocity:

$$\frac{1}{2} m v_p^2 = \frac{1}{2} m v_s^2 + E = \frac{1}{2} m v_s^2 [1+G]$$

where $E = \mu B$

and $G = \frac{\mu B}{k T_s}$

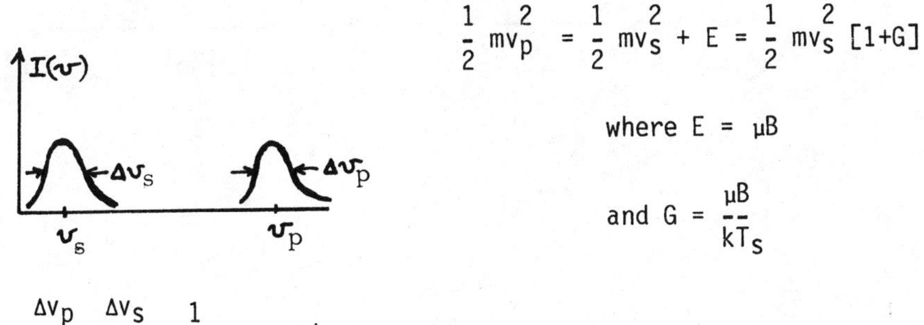

So $\frac{\Delta v_p}{v_p} = \frac{\Delta v_s}{v_s} \frac{1}{[1+G]}$, and

the source is quite monochromatic.

Another slight advantage is that the fringing field of the solenoid magnet will reduce the angular divergence, as it will have a small focusing effect, not sufficient however to compensate for the transverse energy of 1/4 °K. Another problem mentioned, is what happens to the other polarization which will probably not leave the magnet: an appropriate pumping system will have to be designed. Cryogenic pumping should work well.

The role of the acceleration has also been clarified. The higher velocity (E~6K) reduces the ionizer efficiency compared to having the same velocity (E ~ 0.5K), but it allows an excellent focusing factor $f \approx 1$: One can understand this as an admittance matching, and an optimum in beam brightness is probably expected as a function of velocity.

The technical aspects involved were also briefly mentioned, especially the heat load to be expected in the source. The thermalizing heat load is

$$10^{19} \text{ atom/sec} \times 4°K \times 1.4\ 10^{-23} \frac{\text{joule}}{°K} = 5 \cdot 10^{-4} \text{ watt/sec}$$

which is negligible. However, there could be a load due to recombination heat. In summary, the technical aspects do not seem to be particularly difficult.

The advantages obtained in this non conventional approach are summarized in Table 1.

	Conventional Source	Cold Solenoid Source	Gain
Ω	$3 \cdot 10^{-3}$	.5	150
f focusing factor	.3	1	3
throughput*	1	1/25	1/25
ionizer efficiency†	1	7	7

An overall gain factor of 100 appears possible with a cold solenoid source such as proposed by D. Kleppner.

*If density at source remains constant, then throughput $\propto \sqrt{T_{source}}$.

†Ionizer efficiency $\propto 1/\sqrt{T_{ionizer}}$.

Storage

In this discussion the possibility of storing the polarized hydrogen was reviewed: an example for such a scheme is shown as follows:

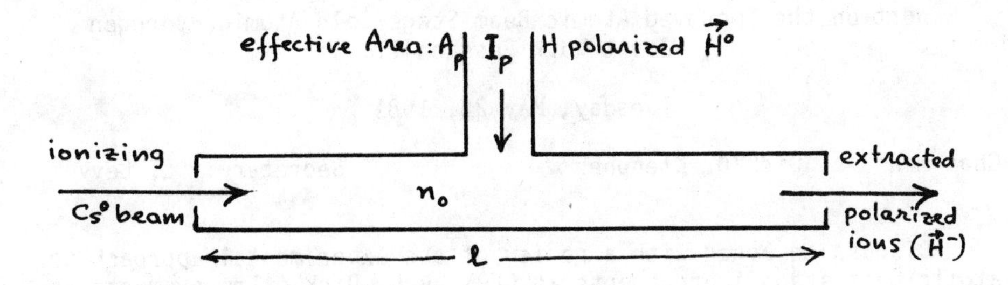

The polarized incoming intensity being

$$I_p = n_p A_p \overline{v}_p$$

leakage intensity

$$I' = \frac{n_o}{4} \overline{v}_o \Sigma A_o$$

leading to a density

$$n_o = \frac{4A_p}{\Sigma A_o} \times \frac{\overline{v}_p}{v_o} \times n_p.$$

The relevant factor for the ionizer efficiency being $n_o \times \ell$, this appears to be also very promising.

Conclusion

Both schemes discussed in this session appear to give large advantages to intensity compared to present sources. Based on these general arguments, it appears that improvements by two orders of magnitude may be possible.

Report on the Improved Atomic Beam Stage/Cold Atomic Hydrogen Joint Work Session III

Tuesday, May 26, 1981

Chairmen: L. Dick/D. Kleppner   Secretary: L. Levy

Discussion began with a review of the experimental approach to atomic beam stage improvements at CERN by L. Dick. Improvements are being planned in the following three areas: 1) Dissociator, 2) Cooling, and 3) Optics. The dissociator bottle will be coated with sapphire which has very good thermal conductivity. In normal use, this will be cooled to 20-30°K. Dissociation will be initiated by microwaves of a few watts. Further cooling to 1-4°K is also planned. A 30 kG solenoid for spin selection can also be used as a storage bottle. A permanent sextupole magnet will also be tried. This can operate at fields of 1.3T which will result, along with more cooling capacity, in a shorter sextupole length.

D. Kleppner then reviewed his ideas for a very cold $\vec{H}°$ source. His first calculations were done to determine the limitations on source density in his proposed sources. Two parameters affect the density: the gas flow, and the ability to carry away heat from recombination reactions. The density limit due to gas flow supply in present systems seems to be $\sim 10^{16}/cm^3$. A reference for recombination processes was given:

Morrow, et al., PRL <u>46</u>, 195 (1981).

The heat load from recombination for various densities is shown in the following table:

$n(cm^{-3})$	P(watts)
10^{16}	$20x10^{-3}$
$3x10^{16}$	.2
10^{17}	2
$3x10^{17}$	20

Therefore, substantial cooling power is needed for densities > $10^{16}/cm^3$. An alternate method was mentioned to reduce the cooling power needed. An intermediate stage (between 4.2°K and 0.5°K) of 1°K could be inserted since less cooling power is needed at higher temperatures. Using this method, the limiting density value is increased to $\sim 6x10^{16}/cm^3$ at a gas flow of 1 torr-liter/sec. Cooling power needed at this density is ~10 watts. A calculation was then done to determine the density of polarized atoms obtainable from a source density of $10^{16}/cm^3$. The density obtained was $n_p = 7x10^{13}/cm^3$. At this density, the output current would be ~100 mA!

One problem addressed was that of the beam particle scattering due to high beam densities. V. Hughes gave references on this:

Gersh and Bernstein, Chem. Phys. Lett. Vol. 4, 221 (1969), and also two references from London-Browning and Buckingham.

The total scattering cross-section for H→H seems to be very small for the velocities projected for a source of this type. Therefore, beam particle scattering should not pose a serious problem.

The last topic discussed was the concept of storage of $\vec{H}^\circ$ in a cooled magnetic bottle. The storage time to build up 10^{17} atoms/cm^3 is ~1 sec at 2T. It was then mentioned that if a Cs° beam was put through such a storage bottle at ~5°K, the H^- current output would be enormous. A problem of this method is that the Cs^+ ions produced (also H^-) might destroy the bottle by heating it up, thereby evaporating the He film. Therefore, getting the H^- out of the bottle may destroy the storage mechanism, but for pulsed operations, there should be enough time between pulses to build up sufficient density by re-forming the storage bottle. The extremely cold hydrogen source seems to be a very promising idea that should be investigated further.

INTENSE UNPOLARIZED NEGATIVE ION SOURCES*

TH. SLUYTERS
Brookhaven National Laboratory
Upton, New York, 11973

Significant progress in a specific topic of science and technology is in general triggered by the urgent need for it. Although negative hydrogen ions were always of special theoretical interest, because it is the most simple negative ion one can think about, the early motivation for negative ion source development was doubling the energy of the particles in medium energy or tandem type of accelerators. Then it was their application in high energy accelerators and storage rings, replacing the complicated proton injection schemes for the simpler and more effective multiturn negative ion injection. But the main thrust in negative ion development arrived recently from the need for steady state multiampere negative ion based neutral beams in the energy range of several hundred kiloelectron-volts for injection into advanced fusion devices. Most of the recent progress in the production and neutralization of negative hydrogen ions and beams described in this review can be found in Ref. 1.

Negative hydrogen ions can be produced a) in hydrogen plasmas, b) by particle scattering or desorption from low workfunction surfaces and c) by electron capture processes of protons or neutrals in a charge exchange cell.

The extraction of negative hydrogen ions from plasmas is limited by the large differences in cross sections between the production and destruction processes. Until recently the most likely production process was dissociation recombination ($e+H_2^+ \rightarrow H^+ + H^-$) with a cross section of 5×10^{-18} cm^2 at 0.3 eV while the most likely destruction process ($e+H^- \rightarrow H+2e$) at 10 eV has a cross section of 4×10^{-15} cm^2. In 1978 Allen and Wong[2)] discovered that vibrationally excited molecules for levels beyond v=4 could increase the production cross sections in the dissociative attachment process by four to five orders of magnitude (up to the 10^{-16} cm^2 scale). It was suggested that these excited molecules could easily be formed by the dissociative recombination, $e+H_3^+ \rightarrow H_2$ (v>6)+H, such as in the low energy discharges produced at the Ecole Polytechnique[3)]. Despite these new discoveries of H^-production processes in plasmas, it is unlikely that these types of sources can be developed in the near future into negative ion sources with extraction densities beyond the 1 mA/cm^2 level.

The most dramatic development in negative ion production is the generation of negative ions by particle collisions on alkali coated metal surfaces, discovered in 1971 by the Russian scientist Dudnikov[4)].

*Work done under the auspices of the U.S. Department of Energy.

Negative ion production on cesium covered metal surfaces can be explained by either desorption of H^- ions from these surfaces or by backscattering from the surface. In the desorption process the H^- ions, adsorbed on the surface, are desorbed by energetic Cs or H particles. In the case of backscattered particles energetic H particles are reflected from the cathode, retaining an appreciable fraction of their incident energy and capture electrons to form negative ions on or near the metal surface. The energy analysis of H^- ions from these sources shows a broad spectrum and thereby the complexity of the process. The energy range of the ions in magnetron sources is in general equal to or lower than the cathode voltage value, which suggests a large probability for the desorption process formation of the H^- ions. In hydrogen discharges, the metal surface with low workfunction may be loaded with hydrogen atoms in the negatively charged state. The theoretical analysis is rather complex: both desorption and backscattering yields depend on the workfunction (cesium coverage of the substrate), the angle of incidence, and the energy and mass of the incident particle.

The most promising negative hydrogen surface plasma sources are the magnetron sources (Novosibirsk and Brookhaven), the magnetron sources with independent plasma control (Brookhaven), the self-extraction sources (Berkeley), the Penning sources (Argonne, Brookhaven, Los Alamos and Oak Ridge) and those using the double charge exchange on surfaces (FOM-Institute, Amsterdam).

The basic configuration of the magnetron source is shown in Fig. 1. The discharge is shaped in the race track of the vacuum chamber in an $E \times B$ electro-magnetic field. The cathode is covered with a thin layer of cesium. Positively charged particles from the discharge and reflected energetic atoms from the walls bombard the low work function cathode and create the H^- particles with at least a 10% probability. The reflected ions either survive the plasma sheath or loose their electrons, pass the expansion chamber of the anode and reach the emission slits. The electrons are deflected in the crossed $E \times B$ field and the H^- are extracted. A disadvantage of this type of source is the relatively high background pressure (0.1 Torr) required to sustain the discharge. Only when the source operates in the d.c. mode and at very high arc current densities, gas efficiencies near 20% can be expected. Figure 2 shows the beam current and beam current density as a function of the arc current.

The magnetron is not only a good source for dense negative ion beams, but it is also an excellent source for dense, low-energy neutral beams, which may become useful for polarized sources[5]. The large electron-loss cross section of H^- in plasmas provide a dense neutral beam with an averaged energy of the cathode potential (150 eV), an energy spread of ~ 20-30 eV and a divergence of about 100 mrad.

In order to improve the gas efficiency in the classical negative ion magnetrons, Brookhaven recently introduced a magnetron based on plasma injection from a hollow cathode discharge[6]. From the plasma

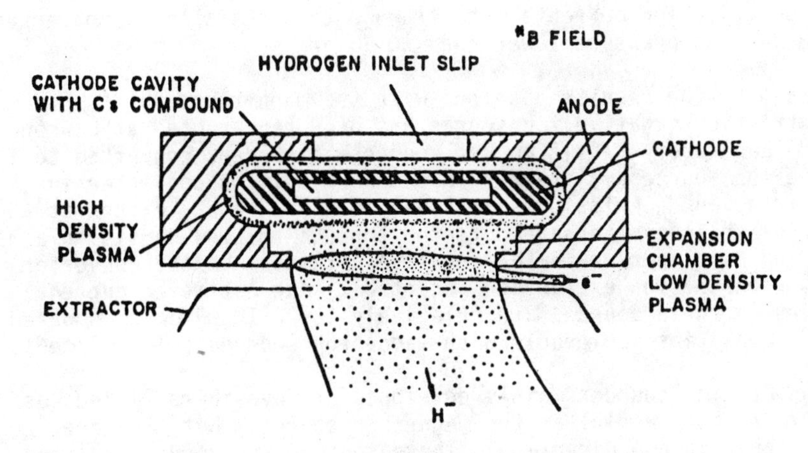

FIGURE 1
Basic configuration of the magnetron negative ion source.

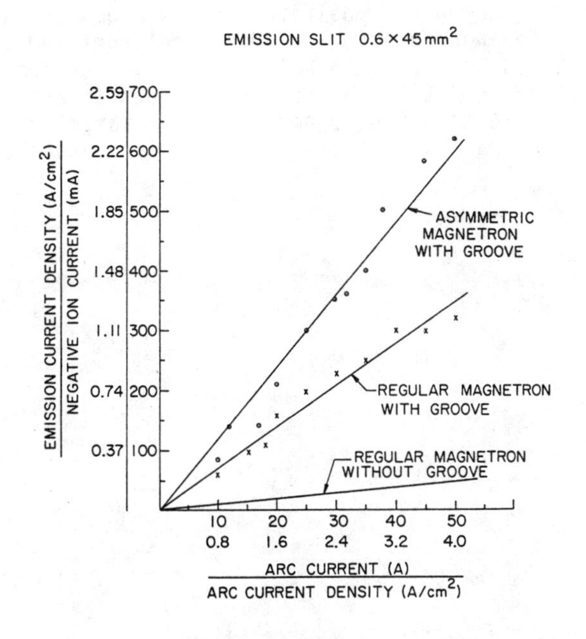

FIGURE 2
Beam currents and beam current densities for the regular magnetron and improved versions.

column positive ion currents were drawn onto a cesiated Mo converter in a background pressure lower than 10^{-3} Torr.

Based on the results obtained with the magnetron, a 1 A self-extraction negative ion source has been constructed at Lawrence Berkeley Laboratory [7]. A bucket source was thereby converted to a negative ion source by inserting a low workfunction converter in front of the anode emission slit (see Fig. 3). Negative ions at the surface of the converter are directed through the low density plasma to the ion exit aperture. The negative potential on the converter becomes then the ion extraction potential. The estimated current densities in these sources are relatively low (~10 mA/cm^2) compared with the densities achievable with magnetron sources (~100 mA/cm^2).

The Penning sources mainly developed at Novosibirsk[5] and Los Alamos[8] are as powerful as the magnetron sources, but they are scalable only in one direction. These sources are very useful for accelerators and they produce beam currents of about 100 mA. Their beam quality (emittance) is in general better than the beam quality from magnetrons.

An interesting combination of positive ion technology and surface ionization is a proposal by a Dutch team of scientists[9]. Instead of converting energetic positive ions in a gas or vapor target, they use a low workfunction solid state target and grazing incidence positive ion beams. With half a monolayer cesium coverage on a molybdenum target negative ion reflection coefficients of 40% are reported. It is interesting to note that the target thickness and the optimum energy are practically the same parameters used in

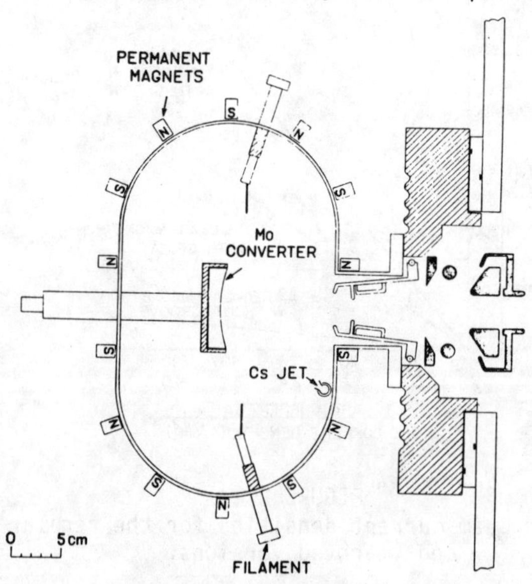

FIGURE 3
The Berkeley multi-cusp self-extraction negative ion source.

double charge exchange in vapor targets.

Finally the third method of negative ion production is based on charge exchange of hydrogen in a suitable gas or vapor target. These targets are in general alkali metals and negative ions of hydrogen have been produced in cesium and sodium vapors as large as 1-2 amperes[10]. The fundamental processes in the formation of these beams are well understood and were predicted to within 20%. The charge exchange method has been applied most extensively by the designers of negative ion based polarized ion beam systems.

References

1) Proceedings of the Second International Symposium on the Production and Neutralization of Negative Hydrogen Ions and Beams, Oct. 1980, edited by Th. Sluyters (Brookhaven National Laboratory, BNL report 51304).

2) M. Allen and S.F. Wong, Phys. Rev. Letters 41, 1791 (1978).

3) M. Bacal and G.W. Hamilton, Phys. Rev. Letters 42, 1538 (1979).

4) Yu. I. Belchenko, G.I. Dimov, V.G. Dudnikov, Dokl. Akad. Nauk. SSSR 213, 1283 (1973).

5) V.G. Dudnikov, Proceedings of the Sec. Int. Symp. on the Production and Neutralization of Neg. Hydr. Ions and Beams, 1980, p. 137, BNL 51304.

6) A. Hershcovitch and K. Prelec, Ibid, p. 160.

7) K.W. Ehlers and K.N. Leung, Ibid, p. 198.

8) P. Allison, H.V. Smith, Jr., J.D. Sherman, Ibid, p. 171.

9) H.J. Hopman, P.J. van Bommel, P. Massmann, E.H.A. Granneman, Ibid, p. 233.

10) E.B. Hooper, Jr. and P. Poulson, Ibid, p. 247.

Report on the High Intensity Unpolarized Source Work Session I

Thursday, May 21, 1981

Chairman: T. Sluyters Secretary: S.R. Magill

Discussion concentrated on formation of unpolarized beams used to ionize H° (to H^-). The suggested topics for discussion were 1) Collinear colliding beams, 2) H^++H^- neutral beam, and 3) Crossed beams.

1. Collinear colliding beams -Two charge exchange reactions with high cross sections were discussed:

$$\vec{H}^\circ + Cs^\circ \rightarrow \vec{H}^- + Cs^+$$

$$\vec{H}^\circ + D^- \rightarrow \vec{H}^- + D^\circ$$

W. Haeberli described the colliding Cs° beam method used at the University of Wisconsin. The Cs gun delivers a 40 keV, equivalent current density of 3mA/cm^2 C_s° beam to the ionizer region resulting in an H^- beam current of 3μA. P. Schultz described a similar Cs beam being developed at Argonne National Laboratory. The parameters of the Cs^+ beam are 30mA at ~42 keV. After neutralization, the resultant Cs° beam equivalent current will be ~ 25 mA focused ~1 m away to a 1 cm diameter. Improvement in beam current is in part due to pulsed operation and in part due to the use of a solid alumina-silicate Cs^+ ion emitter. It was concluded that the use of Cs beams to produce H^- is an attractive and reliable method.

The attractiveness of the $\vec{H}^\circ$, D^- reaction is due to the increasing cross section with decreasing beam energy characteristic of a resonant charge exchange reaction. It was suggested that the Grenoble D^- source could be used in this configuration.

2. H^++H^- - Another suggestion was to inject a combined positive and negative ion beam into the ionizer. Such a combined beam can be transported a long distance (space charge neutralized) and be used to charge exchange with the H°.

0094-243X/82/80150-05 $3.00

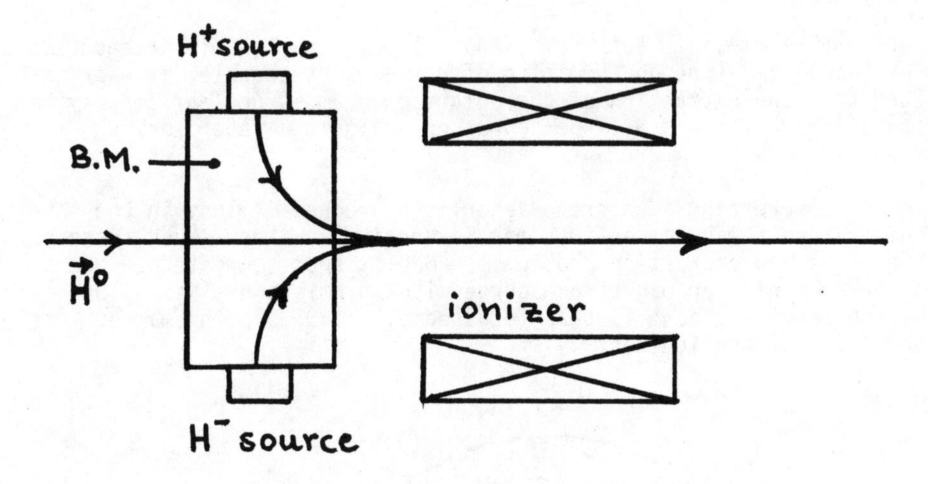

3. Crossed Beams - As an alternative to collinear colliding beam methods, crossed beams were suggested. A method was suggested utilizing a hollow cathode plasma discharge to provide a sheet of D^- beam as shown in the following diagram.

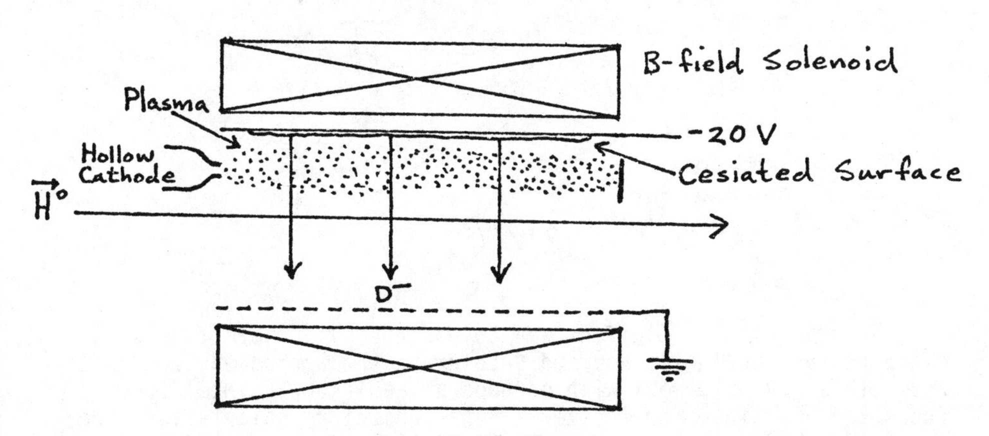

The D^- beam current density from this source is large (~0.1 A/cm^2) and can be made in a sheet of variable width. It was thought that the background plasma pressure might attenuate the H° beam but calculations show that this is not a problem:

$$\text{Attenuation} \propto \exp(-N\sigma\ell)$$

where $N \sim 10^{12}/cm^3$, $\sigma \sim 10^{-16}$ cm^2, and $\ell \sim 10$ cm resulting in attenuation $\propto \exp(-1000)$. Although formation of H^- seems reasonable, it was noted that the extraction of the polarized beam is unclear and may be very difficult. This will be addressed again in a later working group.

An interesting idea from Haeberli to produce D^- ions in the ionizer itself was looked into more extensively without much success (beam loss and extraction problems). Mobley then suggested to use a standard Penning or magnetron source with an emission slit as long as the ionizer. The beam is then focussed by a 180° bending magnet into the center of the ionizer.

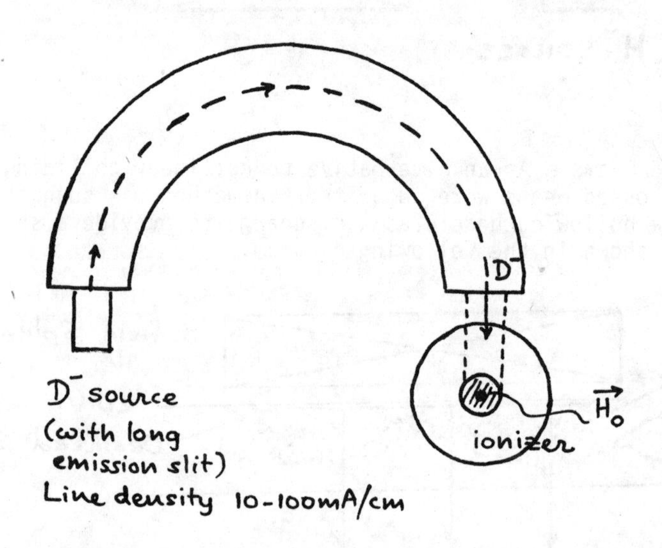

A suggestion by Steffens was to use the H^-/D^- beam from a magnetron or Penning source and decelerate the extracted beam to 2 keV. This may well work with perhaps an additional Einzel lens to focus the beam into the ionizer. Beam trajectory calculations are required. Beam current densities 10-30 mA/cm^2 can easily be achieved.

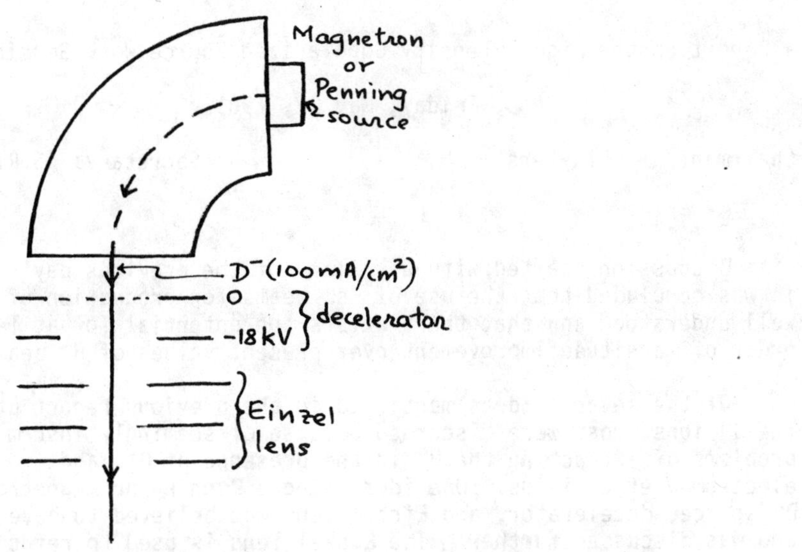

There is the general feeling that the presence of intense unpolarized H^-/D^-sources and beams will help significantly to increase the polarized beam.

Report on the High Intensity Unpolarized Source Work Session II

Friday, May 22, 1981

Chairman: T. Sluyters Secretary: S.R. Magill

Discussion started with a summary of the previous day's ideas. It was concluded that the use of Cs° beams for production of H^- is well understood and that there exists the potential for at least an order of magnitude improvement over present values of H^- beam current.

Of the several ideas mentioned in the previous report utilizing the D^- ions, most were discarded because of seemingly insurmountable problems of extracting the H^- in the presence of D^- and electromagnetic fields. One idea using a Penning or magnetron D^- source, decelerator, and Einzel lens was believed to have potential and was discussed further. The Einzel lens is used to refocus the decelerated D^-beam and the ionization takes place in a region free from large electric fields. The problem of space charge blow-up in the ionizing region still remains and the introduction of a heavy positive ion such as Cs^+ was suggested to help provide space charge neutralization. Since this method may have potential for substantial improvement due to the large reaction cross-section for H_0 on D^-, it was suggested that a comprehensive study, with perhaps a 2-3 man-year experimental effort should be undertaken to investigate this method.

Also a multiple source idea was mentioned as a guaranteed way to increase H^- beam intensity. The resultant beam would have a larger emittance, but it was determined that existing single sources have an emittance of a factor 5-10 lower than the tolerance of linear accelerators. Arising from this idea was a multiple beam idea utilizing permanent magnet sextupoles as shown in the following diagram.

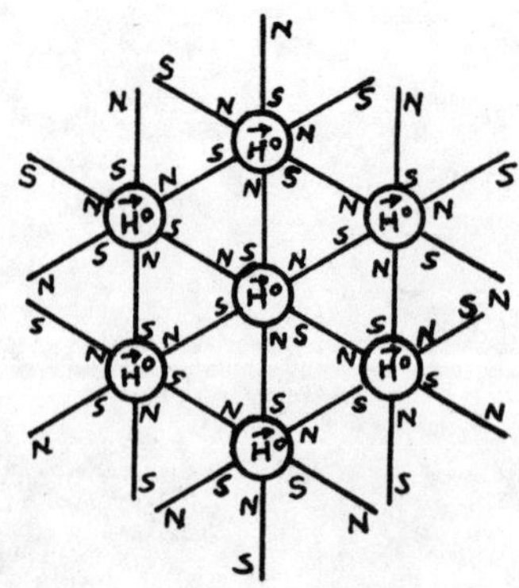

Development of the multiple source seems to be an economic problem and therefore further discussion was deferred.

Possible Uses of Optical Pumping for the Production of Polarized H^- Ions

L. W. Anderson

Department of Physics, University of Wisconsin, Madison, WI 53706

Abstract

The possible use of optical pumping to produce beams of polarized H^- ions is discussed. The proposed H^- ion source involves H^+ ions incident on an optically pumped target where they pick up a spin polarized electron in a high magnetic field. The fast H^o atoms formed pass diabatically through zero field into an oppositely directed field where they are converted into H^- ions in a second target. A critical analysis of the various aspects of the proposed optical pumping charge transfer source is presented.

I. Introduction

In 1957, Zavoiskii[1] proposed the production of polarized ions by the pick up of a polarized electron. In Zavoiskii's proposal a H^+ ion beam is incident on a magnetized ferromagnetic foil. After passing through the foil some of the H^+ ions are neutralized by the pick up of a polarized electron and emerge as fast H^o atoms. Because of the hyperfine interaction the fast H^o atoms have a non-zero time average of the nuclear polarization. Zavoiskii proposed that the fast H^o atoms be ionized in a second foil to form either polarized H^+ ions or polarized H^- ions. Haeberli[2] pointed out that some of the problems associated with the charge transfer production of polarized ions would be solved if the ferromagnetic foil were replaced by an electron spin polarized target such as a polarized stored H^o target or an optically pumped alkali target. The use of optical pumping to produce polarized H^- has been analyzed by Anderson.[3] The current interest in the use of an optically pumped alkali target for the production of polarized H^- ions arises because intense cw dye lasers at wavelengths suitable for optically pumping Li, Na, K, or Rb are now commercially available. Fig. 1 shows a schematic diagram of the

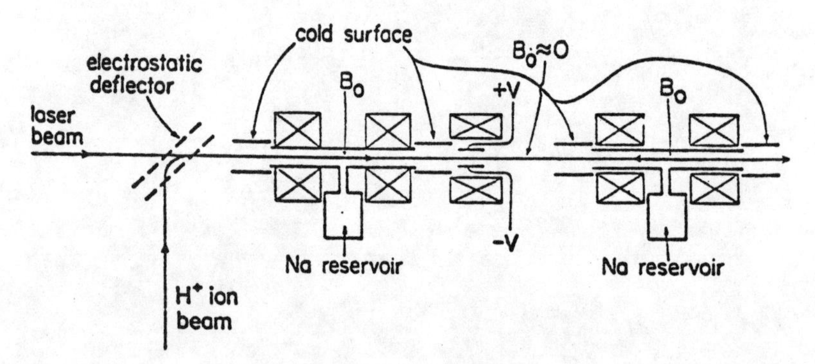

Fig. 1. Schematic of the proposed optically pumped polarized H^- ion source.

optically pumped H^- ion source proposed by Anderson.[3] A beam of H^+ ions is incident on the first Na target with an energy of 5 keV. This target, which is in a large magnetic field, is electron spin polarized by optical pumping with a dye laser beam. In the first target some of the H^+ ions are neutralized by the reaction $H^+ + \vec{Na} \rightarrow \vec{H}^0 + Na^+$ where the arrows indicate that the polarized electron of the Na target is transferred to the H^0 atom that is produced in the reaction. The fast H^0 atom emerges from the first target and enters the second target. The magnetic field at the second target is directed opposite to the field in the first target. The fast electron spin polarized H^0 atoms pass diabatically through a region of near zero field between the two targets. This diabatic passage through zero field transfers the electron spin polarization to the nuclear spin.[4] Some of the fast nuclear spin polarized H^0 atoms are converted to polarized H^- ions in the second target.

The interest in an optically pumped H^- ion source can be understood from the following crude estimate of the potential H^- ion current.[3] A 1W dye laser operating at a wavelength of 589.6 nm produces 3×10^{18} photons/sec. For a Na atom in a high magnetic field (I decoupled from S in the 3s level) 1.5 photons are required on the average to polarize the Na atom. Thus 2×10^{18} Na atoms can be polarized per sec. At a target temperature of 600°K the average time between wall collisions is about 10^{-5} sec assuming the target is a long tube about 1 cm in diameter. Thus even if the Na atoms are completely depolarized at each wall collision a target with $\pi = 10^{13}$ atoms/cm^2 is possible. The long target assures that little polarization is lost by effusion out the ends of the tube. The charge transfer cross section for the reaction $H^+ + Na \rightarrow H^o + Na^+$ is 6×10^{-15} cm^2 at 5 keV[5] so that about 6% of the incident H^+ ion beam picks up a polarized electron. Since the equilibrium fraction of H^- ions emerging from the second Na target is 7.3% at 5 keV[5] about 4×10^{-3} of the incident H^+ beam emerges from the second target as polarized H^- ions. Thus a current of about 4 μA of polarized H^- ions per mA of incident H^+ ions is expected. The technical problems involved in producing polarized H^- ions by charge transfer in an optically pumped target are discussed in this paper.

II. Polarization of H^- Ions from a Charge Transfer Source.

Assume the Na vapor in the first target has an electron spin polarization, P_e. If the H^o atoms produced in the reaction $H^+ + \vec{Na}^o \rightarrow \vec{H}^o + Na^+$ were formed entirely in the ground level then the electron spin polarization of the H^o atoms would be P_e. Since the first target is in a high field (B>507G) the electron spin and the nuclear spin of the H^o atom are decoupled and the nuclear polarization $P_n = 0$. The sudden field reversal between the two targets followed by electron pick up in a high field (B>507G) will produce H^- ions with nuclear polarization $P_n = P_e$.[4] Unfortunately the reaction $H^+ + \vec{Na}^o \rightarrow \vec{H}^o + Na^+$ leads to the H^o atoms being produced primarily in the $n = 2$ level rather than in the $n = 1$ ground level. The electron spin polarization of the H^o atom is partly lost in the radiative transitions that lead to the ground level. If the electron pick up is into the 2p level then the polarization of the ground level atoms following the $2p \rightarrow 1s$ radiative transition is only about 0.41 P_e. Electron capture into the 2s level will result in a similar polarization loss since this level decays primarily by electric field mixing with the 2p level. Pick up into levels higher than $n = 2$ results in cascading and an even larger loss of polarization. Although the exact polarization that can be achieved can not be calculated without detailed knowledge of the cross sections to produce the H^o atoms in various levels it is clear that radiative transitions that lead to ground level H^o atoms result in the loss of substantial electron spin polarization and hence ultimately the loss of nuclear polarization. The loss of electron spin polarization due to radiation can be avoided by the use at the first target of a magnetic field sufficiently strong to decouple L and S in the 2p level of H. Fig. 2 shows, as a function of the magnetic field[6], the electron spin

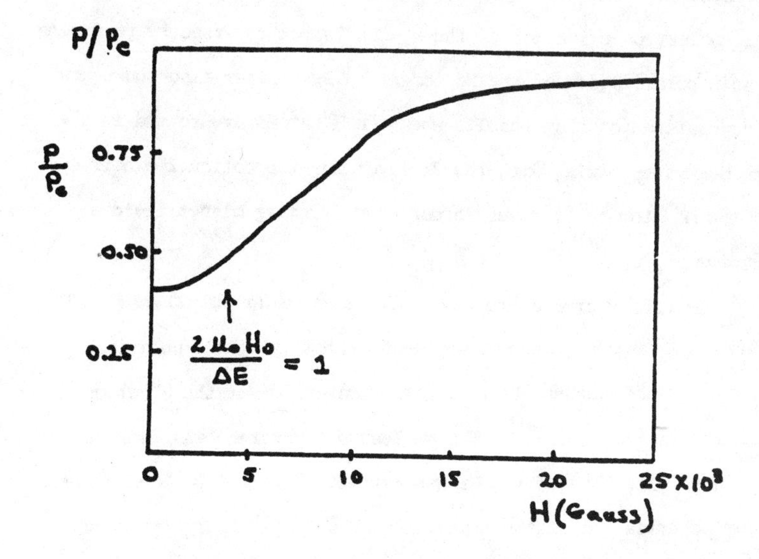

Fig. 2. P/P_e vs H_o.

polarization of the ground level H^o atom that is formed if an electron of polarization P_e is picked up in the 2p level followed by radiative decay to the ground level. A field as large as 10 kG or more is needed to prevent the loss of a substantial fraction of the electron spin polarization in the radiative decay. A field as large as 10 kG at the target introduces severe problems for the ion optics of the neutralization process as discussed in Sec. V of this paper.

In the calculation of P/P_e as a function of H (as shown in Fig. 2) it was assumed that the collision producing the H^o atom in the 2p level results in equal populations of m_L = 1,0, and -1. Hinds et al.[7] have calculated P/P_e as a function of H including the possibility that the collision producing the H^o atom in the 2p level may produce different populations of $m_L = 0$ than m_L = 1 or -1. An interesting result of their calculation is that if $m_L = 0$ is populated differently than m_L = 1 or -1 in the reaction $H^+ + Na^o \rightarrow H^o + Na^+$ then the final nuclear polarization of the H^- ion is not reversed by reversing the electron spin polarization of the Na target. Their paper also shows how to treat the depolarization that results when the H^o atoms are formed in the 2s level which decays by mixing with the 2p level due the motional electric field seen by the H^o atom as it moves through the fringing magnet field at the first Na target.

Witteveen[8] has constructed a prototype charge transfer polarized H^- ion source as follows. A fast H^+ ion beam was neutralized using the polarized Na atom beam from a 6-pole magnet as a target. Conversion of the H^o atoms to H^- ions occurred in a weak field. The nuclear polarization was 14 ± 4%. The electron spin polarization of the Na was estimated to be 0.8. The maximum possible nuclear polarization if the reaction $H^+ + \vec{Na}^o \rightarrow \vec{H}^o + Na^+$ produced only ground level H^o atoms is 0.4. Witteveen attributed the difference between 0.14 and 0.4 entirely to charge transfer from unpolarized background gas. It may be that the difference is largely due to polarization loss in radiative transitions to the ground level.

III. Optical Pumping of a Na Target.

A Na vapor can be polarized by the absorption of circularly polarized

D_1 radiation (the $3^2S_{1/2} \rightarrow 3^2P_{1/2}$ transition). In high field, the Na vapor can be polarized by the absorption of light by the $m_S = -1/2$ levels but not the $m_S = 1/2$ levels even if the light is not circularly polarized.

The hyperfine-Zeeman energy states of Na are shown in Fig. 3. For ^{23}Na the nuclear spin is $I = 3/2$. In a large magnetic field the ground level of Na has 8 sublevels corresponding to the various values of m_S and m_I. The zero field hyperfine separation in Na is $\Delta\nu_{HFS} = 1772$ MHz. The Doppler width for the $3^2S_{1/2} \rightarrow 3^2P_{1/2}$ absorption in Na is about $\Delta\nu_D = 1.7 \times 10^9$ Hz at 600°K. Each of the 4 sublevels with $m_S = -1/2$, $m_I = -3/2, -1/2, 1/2$, or $3/2$ is separated from the next nearest sub level by $\Delta = \Delta\nu_{HFS}/4 = 443$ MHz. In order to optically pump Na in a high field it is necessary for the laser to interact with Na atoms in all 4 sublevels with $m_S = -1/2$, m_I and with a Doppler profile for each level. Thus the laser must interact with atoms having a distribution of absorption frequencies with a width of about $\Delta\nu = (3/4)\Delta\nu_{HFS} + \Delta\nu_D = 3 \times 10^9$ Hz. There are several possible methods of obtaining frequency coverage so that all the atoms in a bandwidth of $\Delta\nu = 3 \times 10^9$ Hz are optically pumped.

(i) The use of high intensity single mode lasers

A laser burns a hole in an inhomogeneous line. The width of the hole is $\Delta\nu_{hole} = \Delta\nu_{homo}\sqrt{1 + I/I_S}$ where $\Delta\nu_{homo}$ is the width of the homogeneous packet, I is the laser intensity, and I_S is the saturation intensity. For Na, $\Delta\nu_{homo} = \frac{1}{2\pi\tau} = 10^7$ Hz where τ is the radiative lifetime of the 3p level. For an optical pumping situation Feld et al.[9] have shown that the saturation intensity is given by

$$I_S = \frac{\hbar\omega}{\sigma\tau}\,\frac{(1 + \tau/T)}{(2 + \Gamma.T)}$$

where ω is the angular frequency of the transition, σ is the optical absorption

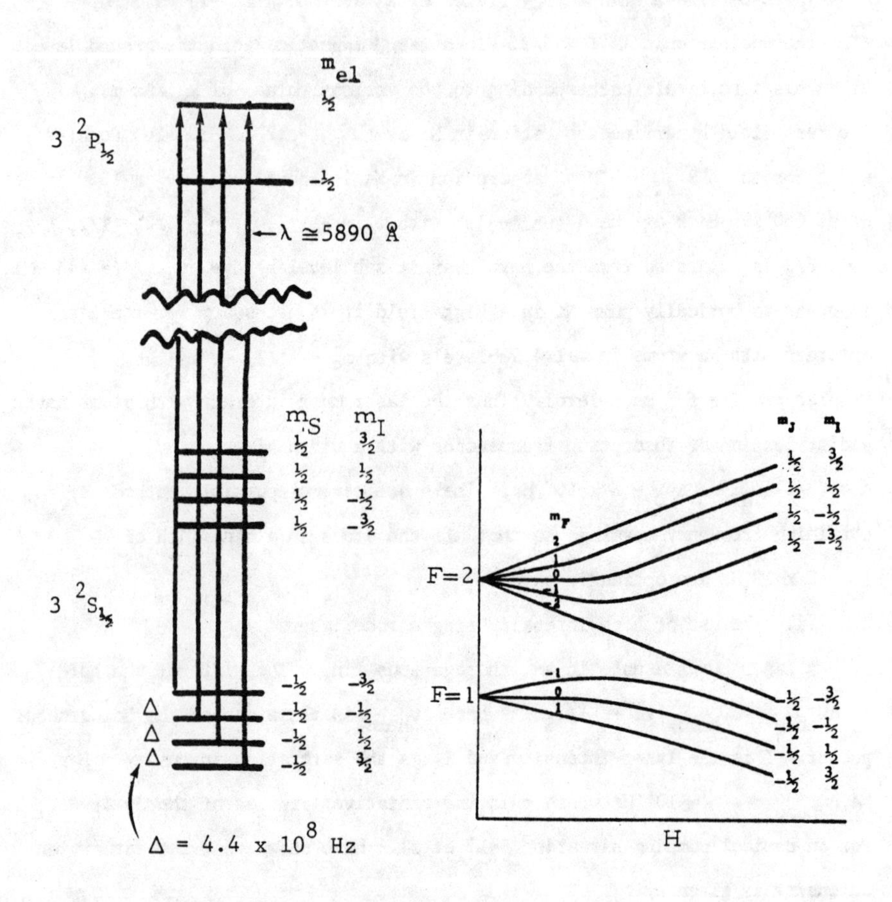

Fig. 3. Relevant energy levels of Na.

cross section at line center for a homogeneous packet, T is the radiation interaction time, and Γ is the decay rate from the excited level into the non-absorbing level. For D_1 pumping of Na in a large magnetic field using circular polarized light, $\Gamma = 1/3\tau$. For D_1 pumping using circularly polarized light, $\sigma = \lambda^2/\pi$. We take T as the time for a Na atom to transverse a 1 cm diameter tube so that $T = 10^{-5}$s. For this case $I_S = 10^{-4}$ W/cm^2. If a 2W single mode laser is used for the optical pumping then $\Delta\nu_{hole} = 1.6 \times 10^9$ Hz. By using 3-4 ring dye lasers operating at (1-2) W power it is possible to optically pump with good frequency coverage a region with a bandwidth of $(3\text{-}4) \times 10^9$ Hz.

A single mode laser can be used together with opticacoustic modulators to generate several optical frequencies. If these optical frequencies are spaced properly it may be possible to obtain good frequency coverage using only one ring dye laser.

Cornelius et al.[10] have optically pumped a Na vapor target using a single mode ring dye laser with an output power of about 400 mW. They use the Faraday effect to measure the average target polarization. They report an electron spin polarization greater than 40%.

It may be that for pulsed ion sources a very high intensity pulsed tunable laser such as the Alexandrite laser can be used for the optical pumping. The Alexandrite laser operates at a wavelength that is suitable for pumping a Rb vapor (but not a Na vapor).

(ii) Use of velocity changing collisions

If an atom makes a number of velocity changing collisions before it hits the wall of the target then the atom has a good chance during part of its radiation interaction time to have a velocity such that the Doppler shifted frequency of the atom is close enough to the laser frequency that the atom can absorb light from the single mode laser. P. G. Pappas et al.[11] have studied the optical pumping of Na in the presence of an Ar buffer gas using a single mode laser. Their results show that the population of the F = 2,m = 2 sublevel of Na depends on the product of the laser intensity I, times the buffer gas pressure, p. For values of $I \cdot p \gtrsim 50$ (mW Torr/cm^2) the population of the F =2,m = 2 sublevel is near 1. This result is valid provided the mean free path of the Na in the Ar buffer gas is much less than the diameter of the Na target.

The problem with the use of a buffer gas is that some of the incident protons are neutralized by pick up of an unpolarized electron from the buffer gas atoms. Let us assume that $I \cdot p = 50$ mW Torr/cm^2 is desired and that $I = 10^3$ mW/cm^2. The Ar pressure must be 5×10^{-2} Torr. This corresponds to a density $\rho \sim 2 \times 10^{15}$ atoms/cm^3 and a target thickness $\pi = \rho \ell \sim 2 \times 10^{16}$ atoms/cm^2. The cross section σ_{+o} for $H^+ + Ar \rightarrow H^o + Ar^+$ is 1.5×10^{-15} cm^2 at 5 keV. Thus $\pi\sigma_{+o} \sim 30$. This is too large. Helium would be a better buffer gas since $\sigma_{+o} = 3 \times 10^{-17}$ for $H^+ + He \rightarrow H^o + He^+$ at 5 keV. However, even with a He buffer gas $\pi\sigma_{+o} \sim 0.6$ if the partial pressure of He is 5×10^{-2} Torr. Again this is too large to be useful. It is clear that it will be difficult to find a buffer gas that is suitable for use in the Na target.

Murnich[12] has suggested that perhaps the alkali density can be increased until alkali-alkali velocity changing collision provide adequate frequency coverage. Murnich[12] has also suggested optically pumping one alkali and using another alkali to provide velocity changing collisions, the second alkali also being polarized by spin exchange collisions.

(iii) Use of a mode locked laser.

A mode locked laser can be used to obtain frequency coverage. A mode locked laser has an output that consists of a train of pulses. The train of pulses is Fourier analyzed in terms of the mode frequencies of the laser, each mode having a particular phase with respect to the other modes. In a Spectra Physics mode locked laser the modes are separated by 80 MHz. If each pulse has a duration of 2.5×10^{-10}s then the Fourier analysis of the light pulse requires an important contribution from the 50 modes in a bandwidth of 4×10^{9} Hz and centered at the Na absorption frequency. Good frequency coverage is obtained if each mode has enough intensity that it burns a hole of width 80 MHz. This requires 5×10^{-3} W per mode in a 1 cm diameter beam. Thus one must have a total power of 250 mW. If one allows for 50% absorption 500 mW power at the entrance to the Na target is required so that one has 250 mW at the exit. It may be possible to obtain this power from one or more mode locked lasers.

(iv) Use of a multimode laser

Both Spectra Physics and Coherent Radiation sell multimode dye lasers with a bandwidth of 4×10^{10} Hz. This can be narrowed to 10^{10} Hz by the use of an etalon. The mode separation in a Spectra Physics laser is about 400 MHz. Thus there are 100 modes in the 4×10^{10} Hz bandwidth. The basic problem with using a multimode laser is that all 100 modes do not lase simultaneously. Instead at a given time only one or several modes lase. The modes that are lasing change in time so a given mode lases only part of the time. The Doppler width of the Na transition is 1.7×10^{9} Hz which is less than the bandwidth of the laser. Hence even when the laser is tuned to cover the Na transition the laser modes that interact with the Na transition lase only part of the time. Thus the Na atom polarization will be a function of the time, being near zero for that part of the time when the laser modes that interact with the Na atoms are not lasing.

We (J. Cusma, D. Swenson, and myself) have studied the optical pumping of a beam of Na atoms with a multimode laser. The polarization of the Na atom beam is analyzed using a 6-pole magnet in a manner that is similar to that used by Baum et al.[13] Using a multimode dye laser with a wavelength of 589.6 nm with a bandwidth of 4×10^{10} Hz and with 0.5 - 1.0 W in a 0.5 cm diameter beam we have pumped a Na atom beam to obtain an electron spin polarization of $P_e \simeq 0.5$. We estimate using an analysis similar to that of Baum et al.[13] that the occupation probabilities of the sublevels are as follows: 0.5 in the F = 2, M = 2 sublevel; 0.18 in F = 2, M = 1 sublevel; 0.12 in the F = 1, M = 1 sublevel; and 0.2 in the remaining 5 sublevels. Using a spectrum analyzer we have observed that the lasing modes of our laser change on a time scale that is less than $\sim 10^{-3}$s. The values of P_e and the sublevel occupation probabilities represent time average values. If one were to use several multimode lasers one might achieve a situation such that most of the time several of the laser modes that interact with the Na would be lasing. Under this condition the average value of P_e would be larger than with only one laser and the time variation of P_e would be smoothed out.

Mori et al.[14] have constructed an optically pumped polarized H^- ion source using a multimode laser for the optical pumping. They monitor the electron spin polarization of the Na target using a 6-pole magnet. They estimate that the electron spin polarization is about 0.7. Mori et al.[14] have obtained a H^- ion current of 4 μA with a 7 keV H^+ ion beam incident and with a magnetic field at the Na target of 10^3 Gauss. They find that the H^- ion current falls to 1 μA as the magnetic field at the Na target increases to 5×10^3 Gauss. Mori et al.[14] have not measured the nuclear polarization of the H^- ion current but they estimate that it is about $\sim$0.3.

The optical pumping of Na by absorption of left circularly polarized $3^2S_{1/2} \rightarrow 3^2P_{1/2}$ radiation from several single mode lasers is described by the following rate equations:

$$\frac{dp_{-1/2}}{dt} = -\beta_o p_{-1/2} + \frac{2}{3}\beta_o p_{-1/2} - \frac{(p_{-1/2} - 1/2)}{T}$$

and

$$\frac{dp_{1/2}}{dt} = \frac{1}{3}\beta_o p_{-1/2} - \frac{(p_{1/2} - 1/2)}{T} .$$

In the above equation $p_{1/2}$ and $p_{-1/2}$ are the occupation probabilities of the $3^2S_{1/2}$ sublevels with $m_S = 1/2$ and $m_S = -1/2$. Thus $p_{1/2} + p_{-1/2} = 1$. The quantity β_o is given by $\beta_o = \sum_i \int I_{oi}\sigma(\nu_{Li} - \nu_o)\, h(\nu_o)\, d\nu_o$ where I_{oi} is the intensity of the ith laser in photons per cm^2 per sec, ν_{Li} is the frequency of the ith laser, ν_o is the atomic absorption frequency, and $h(\nu_o)$ is the inhomogeneous distribution of atomic absorption frequencies. The sum is over the various laser frequencies. The relaxation time T is taken as the average time between wall collisions $T \sim 10^{-5}$s. The equations contain the radiative transition probability of 2/3 and 1/3 respectively for the $^2P_{1/2}$ $m_J = 1/2$ sublevel to decay to the $m_J = 1/2$ and $m_J = 1/2$ sublevels of the ground level. The quantity β_o is given approximately by $\beta_o = \pi r_e f I_o/\Delta\nu$ where $I_o = \sum_i I_{oi}$, r_e is the classical radius of the electron, c is the speed of light and f = 1/6 is the oscillator strength for the absorption of circularly polarized light by the $3^2S_{1/2} \rightarrow 3^2P_{1/2}$ transition in Na. In this expression for β_o we have assumed that $h(\nu_o)$ is very broad compared to $\Delta\nu_{homo}$. The steady state solution to the rate equation is

$$p_{-1/2} = \frac{1}{2(1+\frac{\beta_o T}{3})} = 1 - p_{1/2}$$

This leads to an electron spin polarization of

$$P_e = p_{1/2} - p_{-1/2} = 1 - \frac{1}{(1+\frac{\beta_o T}{3})} .$$

For the case of optical pumping by four 1.0W ring dye lasers $\beta_o = 2 \times 10^7$ sec^{-1} so that $\beta_o T = 200$ and $P_e = 0.99$ at the entrance to the Na target. If 0.50 of the photons are absorbed then at the end of the target $\beta_o = 1 \times 10^7$ and $P_e = 0.97$. We estimate that a target with $\pi = (2\text{-}4) \times 10^{13}$ atom/cm^2 can be produced (see the final paragraph of this section). This corresponds to about (8-16)μA of polarized H^- ions per mA of H^+ incident. Other optical pumping schemes such as using a buffer gas, a mode locked dye laser or a multimode laser give somewhat different results for P_e and π. Experiments on an optically pumped Na target to determine the possible P_e and π for various lasers and targets are needed.

We have assumed that the spin depolarization time T corresponds to the average time between sucessive wall collisions for a Na atom. If the atoms are not completely disoriented at every wall collision T will be longer. Measurements are needed to determine T for a Na target made of Cu or stainless steel. Also a longer value of T may be possible using polyethylene or some other material that prevents depolarization when the Na bounces off the wall. Experiments on wall coatings are needed.

Imprisonment of the resonance radiation can limit the target density. Radiation trapping becomes important when the mean free path of a photon in the vapor is comparable to the radius of the Na target. This corresponds to a Na atom density of $n=1/(\sigma R)$ where σ is the photoabsorption cross section and R is the radius of the Na target. The D_1 absorption cross section at line center in a Na vapor at a temperature of 600°K is $\sigma \simeq 10^{-12}$ cm^2. For R = 0.5 cm this leads to $n = 2 \times 10^{12}$ atoms/cm^3. Imprisonment of the resonance radiation does not result in depolarization. However when imprisonment of the resonance radiation occurs more than 1.5 photons per Na atoms are required to polarize the Na. We estimate with 4 W of laser light and assuming $T = 10^{-5}$s that the maximum Na density is $(2\text{-}4) \times 10^{12}$ atoms/cm^3. This corresponds to $\pi = (2\text{-}4) \times 10^{13}$ atoms/cm^2 for a 20 cm long target.

IV. Ion Beam Current

In Sec. III we concluded that an electron spin polarized Na target with a density such that (8-16) μA of H^- per mA of H^+ incident is possible. The question remains as to how large an incident H^+ ion beam can be passed through a 20 cm long 1 cm diameter Na vapor target? The space charge limit for an H^+ beam is $I_{sc} = 38 \times 10^{-6} (\frac{d}{\ell})^2 (\frac{V^{3/2}}{m^{1/2}}) = 0.8$ mA. Of course space charge neutralization will occur so that currents larger than I_{sc} are possible. Experiments are needed to determine the possible incident H^+ currents. York and Cornelius[15] believe, based on their experience, that currents of 10-20 mA of H^+ through the Na target are possible.

V. The Magnetic Field Problem

If a parallel beam of H^+ ions enters the Na target and neutralizes at the center then a particle in the neutral beam acquires a transverse velocity v_t, given by

$$\frac{v_t}{v_z} = \frac{eB_o r}{2mv_z}$$

where r is the radius of the beam, B_o is the magnetic field, m is the mass of the proton and v_z is the incident velocity.[16] For $B_o = 10^4$ gauss and $r = 0.5$ cm we calculate $v_t/v_z = 0.25$. Clearly $v_t/v_z = 0.25$ is unacceptable since most of the beam after neutralization will strike the Na target tube. Thus one must either operate in a lower magnetic field and accept the lower H^- ion polarization or operate the ion source and Na target in a collinear geometry and in the same large magnetic field. Experiments are needed to determine if either of these is acceptable.

VI. Conclusions

I conclude that there are at the present time too many unknowns to evaluate the ultimate prospects for an optically pumped ion source. The best method of optical pumping to obtain frequency coverage of the Doppler absorption profile of the optically pumped target is a major problem that needs study. The problems associated with the Na target density i.e. the problem of disorientation of the Na due to wall collisions and the problem of radiation trapping need study. Whether an ion source can be successfully run in a field of 10^4 gauss with a small value of v_t/v_z needs study. How much current can be passed through the electron spin polarized Na target needs study.

In addition, there are no rf transitions in the proposed optically pumped ion source so that the source will have limited usefulness for the production of polarized D^- ions.

Acknowledgments

I acknowledge many helpful discussions with Prof. W. Haeberli. I also acknowledge discussions with J. Cusma, D. Swenson, Prof. J. E. Lawler, Dr. D. E. Murnich, Prof. T. W. Hänsch, Dr. R. L. York and Dr. W. D. Cornelius.

Bibliography

1. E. K. Zavoiskii, Soviet Physics JEPT 5, 338 (1975).

2. W. Haeberli, Proceedings 2nd Int. Symp. Polarization Phenomena in Nuclear Recations (P. Huber and H. Schopper, eds.) Birkhauser Basel 1966, p. 64.

3. L. W. Anderson, Nucl. Instr. and Methods 167, 363 (1979).

4. P. G. Sona, Energia Nucl. 14, 295 (1970).

5. C. J. Anderson, A. M. Howald and L. W. Anderson, Nucl. Instr. and Methods 165, 583 (1979).

6. After I calculated the variation of P/P_e as a function of H I received three papers that show the same results: Y. Mori, K. Ito, A. Tagagi, and S. Furkumoto, Bull. Am. Phys. Soc. 26, 129 (1981); E. F. Parker, Bull. Am. Phys. Soc. 26, 126 (1981); and E. A. Hinds, W. D. Cornelius and R. L. York, to be published.

7. E. A. Hinds, W. D. Cornelius, and R. L. York, to be published.

8. G. J. Witteveen, Nucl. Instr. and Methods 158, 57 (1979).

9. M. S. Feld, M. M. Burns, T. U. Kuhl, P. G. Pappas, and D. E. Murnich, Optics Letters, 5, 79 (1980); see also P. G. Pappas, M. M. Burns, D. D. Hinshelwood, M. S. Feld and D. E. Murnich, Phys. Rev. A 21, 1955 (1980).

10. W. D. Cornelius, R. L. York and E. A. Hinds, reported at the work shop on High Intensity Polarized Proton Ion sources at Ann Arbor, Michigan, (1981).

11. P. G. Pappas, R. L. Forber, W. W. Quivers, Jr., R. R. Dasari, M. S. Feld, and D. Murnich, to be published.

12. D. Murnich, private communication.

13. G. Baum, C. D. Caldwell, and W. Schroder, Appl. Phys. 21, 121 (1980).

14. Y. Mori, K. Ito, A. Takagi, and S. Fukumoto, to be published.

15. R. L. York and W. D. Cornelius, private communication.

16. G. G. Ohlsen, J. L. McKibben, R. R. Stevens, Jr. and G. P. Lawrence, Nucl. Inst. and Methods 73, 45 (1969).

Report on the Optical Pumping PPIS Work Session I

Friday, May 22, 1981

Chairman: L.W. Anderson Secretary: L. Levy

In this session two topics were discussed. The first one was a discussion by Y. Mori of the new H^- polarized source built at KEK. The second one was a discussion of the various techniques to measure polarization of optically pumped alkali vapor. Special attention was given to the technique used by W. Cornelius from Los Alamos Laboratories.

1. Polarized H^- Source

Y. Mori first presented the details of the H^- polarized source. In particular, Dr. Mori presented a new method for optical pumping in a low magnetic field and charge transfer in a high field using an optically pumped beam [not cell].

Dr. Mori also presented a method using crossed E and B fields to avoid the emittance problem when charge transfer occurs in a large field. Some discussions were raised by T. Clegg concerning the limitation imposed by the velocity spread of the ion source.

2. Methods to measure polarization of optically pumped alkali targets:

Dr. Cornelius presented a technique employing the Faraday effect to measure polarization using a single mode laser. The polarizations measured were between 40 and 70% independent of target density up to a target thickness of 3×10^{13} atoms/cm^2. The method was applied down to a target density of 10^{12} atom/cm^2 with sufficient signal strength.

The various techniques to measure polarization were discussed:

a) Faraday Rotation: for densities larger than 10^{12}/cm^2

b) Sextupole magnet

c) Study of saturation curves

d) Use of two lasers (a pump laser and a probe laser) for absorption measurements and rate analysis

e) Using a Fabry-Perot interferometer to measure the frequency distribution of the forward and backward scattered light.

0094-243X/82/80173-07 $3.00

Report on the Optical Pumping PPIS Work Session II

Friday, May 22, 1981

Chairman: L.W. Anderson Secretary: M.A. Cummings

Discussion continued on possible methods of measuring polarization of the optically pumped target. Mentioned earlier were the Fabry-perot (interferometer) method by measuring the frequency distribution for forward and backward scattered light, and the Faraday rotator (for $\pi > 10^{12}$ atoms/cm^2). Considered this afternoon were a probe laser method and possible polarized beam production from an unpolarized target in a sextupole. Other topics considered were:

1) Intensity of beam vs. cell target

2) Different target and buffer gases

3) Depolarization in the n=2 states of hydrogen.

One can use a frequency-varying laser probe to measure polarization of an alkali vapor cell. A small intensity laser is directed into the cell and its frequency varied around some energy where the alkali spins are induced to flip. The plot below is the spectrum of population density vs. energy (frequency) of a typical alkali polarization measurement.

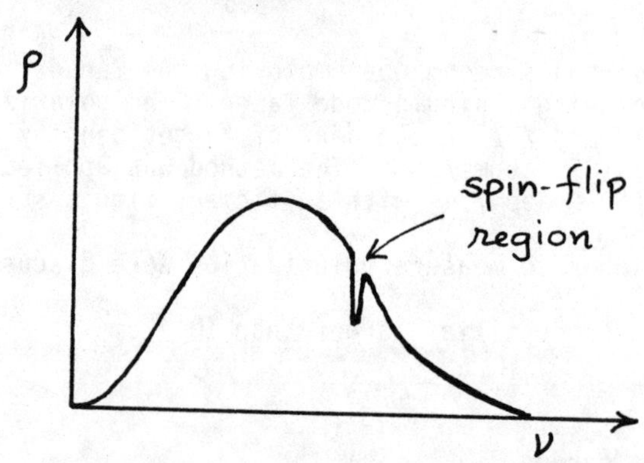

E. Hinds made a calculation for the effective acceptance solid angle in a Na cell. For a laser beam:

$\overline{\Delta\nu} = 2\text{x}10^7$ Hz

$\nu = 5\text{x}10^{14}$ Hz

with fractional spread $\dfrac{\overline{\Delta\nu}}{\nu} = \dfrac{2}{5} \times 10^{-7}$.

The effective Doppler spread of Na vapor is given by

$$\frac{\Delta f}{f} = \frac{V_T}{c} = \frac{3\times10^4\theta}{3\times10^{10}} = 10^{-6}\ \theta, \text{ where } V_T = \text{transverse velocity.}$$

The largest possible θ is at the greatest Doppler spread that can be allowed while still exciting atoms.

$$10^{-6}\ \theta_x = \frac{2}{5} \times 10^{-7} \rightarrow \theta_x = 4\times10^{-2} \text{ rad}$$

and for $\theta_y = .1$ rad, we obtain $\Omega = 4\times10^{-3}$ sr.

This compares closely to the acceptance solid angle of Na beams in a sextupole magnet. However several beams of Na are needed to approach the flux of Na in a vapor cell.

It was suggested that perhaps alkali gases other than sodium could help yield greater beam currents. Lithium for example has cross-sections that peak at beam velocities larger than those of the cross-sectional peaks in sodium. This reduces the space charge density, and therefore as can be shown by using the Langmuir-Child law, increases beam current. There were suggestions for using different buffer gases other than helium to try to increase the radiation trapping limit. There seemed to be no fundamental density limitations involved with increasing beam intensity.

E. Hinds then spoke on depolarizations in the n=2 states of hydrogen. He made the approximation that charge capture is fast in the alkali vapor cell ($\sim10^{-15}$s) so that the spin orientation of the alkali atom is preserved. He used a density matrix to represent the orbital 2P states (which he considered before the 2S state):

$$\rho_{2P}\ (m_s) = \begin{bmatrix} 1-T & 0 & 0 \\ 0 & 1+2T & 0 \\ 0 & 0 & 1-T \end{bmatrix}, \text{ where } T = \sqrt{5\pi}\ \langle Y_{20} \rangle .$$

Cylindrical symmetry of the collision process restricts the possible tensor polarizations to Y_{00} and Y_{20} spherical harmonics. Y_{10} is forbidden by parity.

The time evolution of the density states was a simple unitary transformation:

$$\rho_{2P}(t) = e^{iHt}\ \rho_{2P}(0)\ e^{-iHt}$$

where H is the full Hamiltonian including Zeeman and fine structure interactions. The oscillating off-diagonal terms give rise to quantum beats which oscillate rapidly compared with decay rate to the ground state and average out to zero and the electronic polarization densities may be calculated from time independent diagonal elements of ρ. The polarization density can be expressed as

$$\pi = A(P) + B(T) + C(PT)$$

where P = the spin polarization of the alkali vapor and the constants A,B,C are dependent on the B field.

This expression doesn't violate parity or time reversal, but indicates that the magnetic field breaks symmetry between positively and negatively polarized spins, as seen by the graph below.

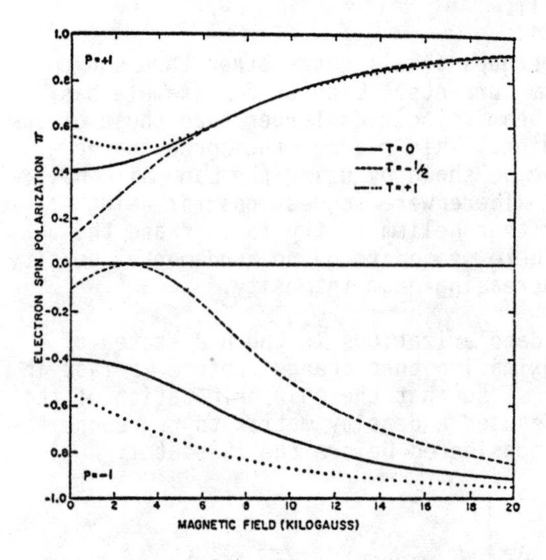

T = 0 implies equally populated m_S states
T = -1/2 (m_S = ±1 only)
T = 1 (m_S = 0 only)

It is apparent that reversing the beam polarization by reversing the alkali polarization is impossible. But it is possible that an unpolarized target can produce a polarized beam.

The quenching of the 2S level by the induced E field (vxB) introduces a mixing of 2S and 2P states and contributes to T = -1/2 in the terms of the π equation.

A large magnetic field is required to maintain polarization which results in substantial increases of the beam emittance. This could be corrected by the ECR principle of having both the ion source and the charge-exchange target in the same magnetic field.

Report on the Optical Pumping PPIS Work Session III

Monday, May 25, 1981

Chairman: P. Schmor Secretary: M.A. Cummings

This afternoon's discussions were primarily about the proposed optically pumped ion source at KEK in Japan.

E. Steffans started with a calculation of the density of the Na beam that could be expected. He assumes that $\rho_{Na} \cong 10^{12}/cm^3$.

For the $H^+ \rightarrow H^\circ$ conversion:

length of target = 10 cm = t

$(H^+ \rightarrow H^\circ)$ cross sec. = 6×10^{-15} cm^2 = σ_{+o}

$$\rho_{Na} \times \sigma_{+o} \times t = (10^{12}\ cm^{-3})\ (6 \times 10^{-15})\ cm^2\ (10\ cm)$$

$$= 6 \times 10^{-2}$$

$$= 6\%\ H^+ \rightarrow H^\circ \text{ efficiency}$$

and for the entire $\vec{H}^+ \rightarrow \vec{H}^-$ conversion the efficiency is ~ .5%:

1 mA source → 5μA $\vec{H}^-$

10 mA → 50μA

The cross section of the Na beam is a 10x1 cm rectangle. This requires that the laser used for optical pumping be widened by an expander lens to 10 cm. Marburg achieved 90% electron polarization of the Na vapor in this manner which is what is expected at KEK. Density limits, radiation trapping and even formation of molecular sodium presents no significant problem.

E. Steffans gave a brief presentation on the Na recycling oven shown in the following figure. Iron is used on the small beam apertures to avoid the formation of Na droplets. Na consumption is low at ~ 1 gm/day.

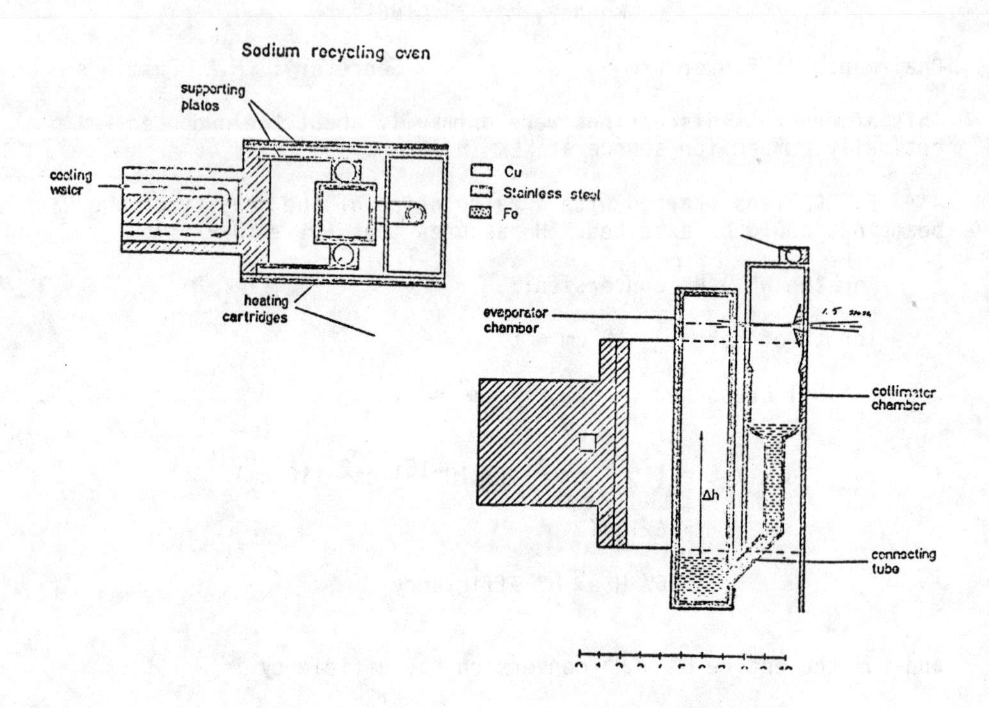

Mori questioned whether or not the cross-sections for (proton capturing e⁻? from) the sodium oven would remain the same in a high B field. Also if any depolarization of the beam would result.

A problem brought up was the unacceptably large emittances in solenoidal B fields. A way out of this problem is the use of the ExB field. This results in good polarization but poor intensity. Therefore the scheme of using several small apertures in a solenoidal magnet in place of one larger hole was considered. The cm divergence pattern of H° for the individual smaller holes is the same as that for the single larger hole but the envelope of their enlarged emittances remains acceptable.

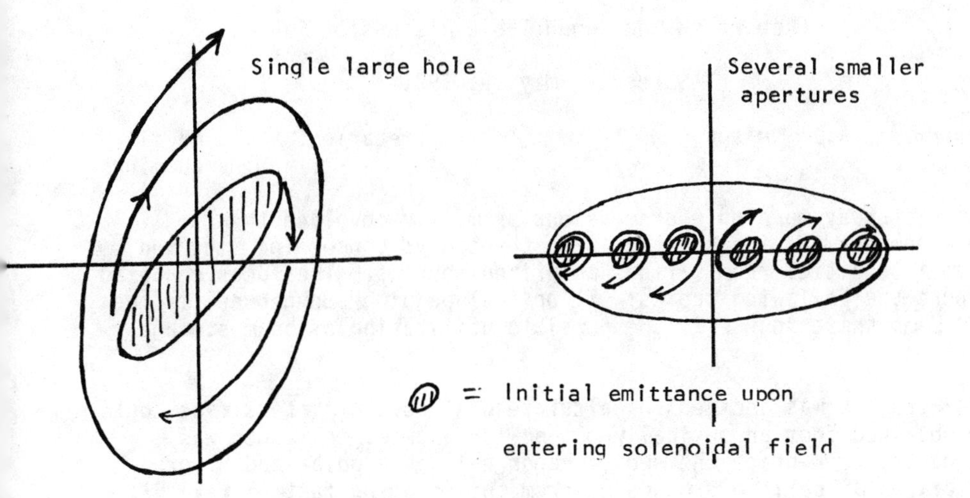

One gets out sufficient current in this manner but it is unclear if one would get out a high e⁻polarization. Also this method does not get rid of the space charge problem. The method utilizing the ExB field effectively overcomes the emittance blow-up problem but probably will not take advantage of space charge neutralization.

Mori noted that one could obtain the polarization of protons as a function of the solenoidal B field without concern of target vapor polarization.

Report on the Round Table Discussion I

Saturday, May 23, 1981

Chairman: A.D. Krisch

Secretaries: S.R. Magill
M.A. Cummings

Saturday morning's discussions primarily involved the developments and ideas of the past three days concerning improved beam intensities from various polarized sources. The focus centered around the following topics: 1) optical pumping techniques, 2) the ECR Lamb shift source 3) the possible utilization of beam storage devices.

1. First it was decided to calculate what sort of intensities could be obtained from an optically pumped ion source. A method was suggested to replace the sodium vapor cell by a polarized H° or polarized D° cell to produce H° from the incoming fast (5 keV) H^+. The 5 keV H° is subsequently transported through an optically pumped sodium cell to produce a polarized H^- beam. This would overcome some of the space charge problem requiring high magnetic fields in the standard $H^+ + D° \rightarrow H° + D^+$ process. This example was used to calculate possible polarized H^- beam intensities. For 5 keV H^+, the cross-section σ on D° is 2×10^{-15} cm^2

$\vec{D}°$ density assumed: 10^{13} cm^{-3}
length of D° cell: 20 cm

$$\text{fraction of } \vec{H}° = (2\times10^{-15}\ \text{cm}^2)\times(10^{13}\ \text{cm}^{-3})\times 20\ \text{cm} = .4$$

Typical efficiency of producing $\vec{H}^-$ from $\vec{H}°$ on Na cell is ~ 0.07. Thus we obtain

$$\frac{\vec{H}^-}{\vec{H}^+ \text{in}} = (.07\times.4) \sim 0.02$$

So, for an H^+ beam intensity of 10 mA, one obtains a polarized H^- intensity of 200 μA. At this rate the loss of D° is 3×10^{16}/sec. It was generally felt that replacement of D° at this rate would present no significant difficulty.

It was also suggested that one could replace the helium buffer gas in the optically pumped vapor cell with rubidium gas. This would provide a denser gas background, increasing the radiation trapping limit.

2. T. Clegg felt that possible intensities from an electron cyclotron resonance (ECR) Lamb-shift source are quite large, but that only a small fraction of the available current could be transported. His idea was to put both the ECR and the cesium cell in a strong solenoidal magnet of 3 kGauss to confine the beam. A problem arises from the huge field gradients at the ends of the magnets, but it was calculated that a gradient of up to 100 gauss/cm could be tolerated without quenching the meta-stable beam. This would limit the beam to a small diameter. Further corrections in target thickness could yield up to a 30% D^+ to meta-stable beam conversion.

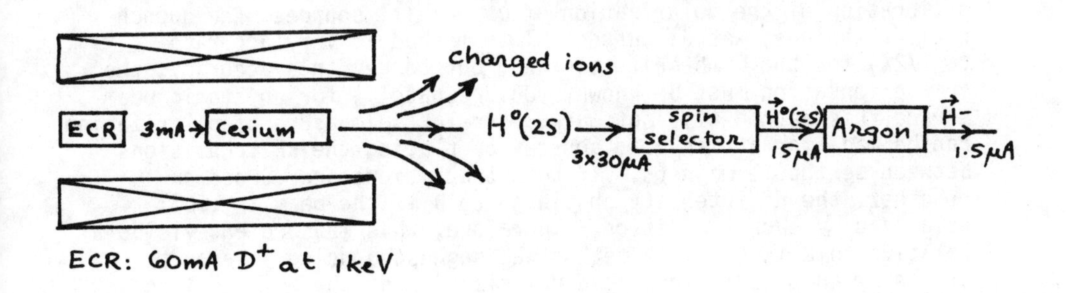

3. Discussion then came to the possible use of a storage ring to increase the beam intensity. It was noted that single turn extraction from a small diameter storage ring would only fill a small fraction of a large accelerator. One would need a slow extraction system from the storage ring to fill the entire large ring. A quick calculation for the AGS accelerator with a storage ring indicated that the storage ring could be filled for 3000 times the present injection time before AGS main ring space charge problems become important. The proposed storage ring for SATURNE, NIMAS, was discussed. Since the storage energy was fairly high the cost was rather high (~23,000,000 FF). Since emittance values of polarized proton sources are generally 5-10 times less than that of unpolarized sources, it was thought that storage devices should be considered in greater detail. If high quality monochromatic beams could be made using the cold front end proposed by Kleppner, storage rings might seem more attractive because of the even smaller emittances.

Report on the Roundtable Discussion II

Monday, May 25, 1981

Chairman: A.D. Krisch   Secretary: S.R. Magill

The first topic addressed was the calibration of the source polarization prior to injection. There are many polarization measurement techniques available at energies above 1 MeV, but it is difficult at the low beam energy of the source. A method used for calibration of the polarization of Lamb-shift sources, the quench ratio technique, was discussed. This method is very accurate (~ 1/2%) for the Lamb-shift sources, but to get this accuracy, the spin orientation must be known. Unfortunately, for an atomic beam source this method will not work. A relative measure of polarization can be made for atomic beam sources by placing the RF transitions between sextupole magnets. If both transitions are turned on together, the H° intensity should go to 0 if the beam is 100% polarized by each transition. Therefore, this measurement yields a relative polarization. A method was suggested using a laser to excite Lyman α radiation. The polarization of the emitted light could be analyzed to determine beam polarization. It was mentioned that the Japanese have worked on a similar method, but the reference was not known.* It was also suggested that Balmer α radiation would give an easier measurement for relative polarization. This method is further complicated if the polarization is a function of radial position in the beam. Therefore, more thinking is needed on this subject.

Another topic discussed was the advantage of a ground voltage pre-accelerator for the source such as an RFQ or Megalac. It was generally agreed that removing the source from a 750 kV dome was very desirable for controls, diagnostics and repairs. It appears that there will be little depolarization with an RFQ preaccelerator since the B-field is 0 on the axis.

The last topic discussed was that of remote operation of diagnostics and controls. It was felt that remote controls were possible for most uses, but some things may be better controlled manually, especially relatively fixed operating controls. It was suggested that more consideration be given in future groups to diagnostics and controls for polarized sources.

----

*S. Jaccard later sent in the information that the Japanese turned out to be Russians, with names like Soroko. The reference is: Lamb-shift polarimeter of the slow Proton Beam, Yu A. Pliss and L.M. Soroko, Nucl. Inst. and Meth. 135 (1976) 497. See also: A proposal for detecting the polarization of slow protons and deuterons, C. Clausnitzer and D. Fick, Nucl. Inst. and Meth. (1967) 171.

Report on the Round Table Discussion III

Wednesday, May 27, 1981

Chairman: A.D. Krisch Secretary: M.A. Cummings

Among the first things brought up for discussion this afternoon was the necessity of more specific fundamental physics research on polarized sources. Such concerns as radiation trapping limits, ionization limits in charged-targets, polarization transitions in laser pumped targets, and polarization measurement techniques should be addressed directly by experimentalists rather than be set aside as mere technical problems. Also, some aspects of ion sources need greater study (dissociators, sextupoles, cooling, etc.). This shift in basic philosophy toward polarized sources would hopefully result in better funding in this area.

Kleppner then started discussion on super-sonic jet targets. The beam diameter problem was brought up. One could open up the aperture of a state selector, for instance, improving collimation; but accelerator people typically feel that beam size should be limited to ~1 cm. A major problem is matching emittances of sources to accelerators. Y.Y. Lee suggested that matching a source to a pre-accelerator could improve emittance matching. It was calculated that LINACS typically have a transverse emittance of ~10 π mm mrad$\cdot$MeV$^{1/2}$.

In smaller accelerators, the problem can be considered in a 6-dimensional phase space. In this case transverse momenta do not have to be very low compared to the longitudinal energy of the beam. J. Arvieux detailed such a phase space argument for an RF transition region in a polarized source. Further noted was that monochromatic beams could be enhanced by optimizing forward velocity but one would get into problems with a large $\Delta v/v$. Concerning atomic beam stages, it was pointed out that after the beam is shunted into the charge exchanger, the phase space is totally altered. It turns out that it is more important to maximize the number of atoms going into the charge exchanger than the forward velocity of the beam, in optimizing brightness to volume ratios.

There was a discussion of making beams as directional as possible. A short calculation of beam through-put was done using a mean free path argument, and it was found that beam transfer at low temperatures compared favorably to using high magnetic fields. Another suggestion was conducting the beam down a tube after state selection.

L. Dick talked about CERN's plans for improvement of the atomic beam stage. One idea was to form a molecular supersonic jet and then run it through a dissociator. Acceleration, however, was considered too fast for this process to work. A more feasible scheme to take at this point is shooting a cold H beam into a polarized jet. A

$10^{17}/cm^3$ intensity is now possible but more study is needed on beam-jet interactions if intensities are to be increased to the desired $10^{20}/cm^3$. A problem to be dealt with is the energy stored in high density beams. Finally the possibility of storing a polarized jet in the beam has yet to be investigated.

Acceleration of Polarized Protons at Saturne: First Results

J. Arvieux*
Laboratoire National Saturne, 91191 Gif- sur-Yvette, France
and Institut des Sciences Nucleaires, 38026 Grenoble, France

The accelerator SATURNE is a synchrotron which accelerates particles up to P/Z = 3.8 GeV/c. Thus the maximum energy for protons T_p is about 3 GeV, and for deuterons T_d is about 2.3 GeV. It is equipped with a polarized ion source (HYPERION, the name of a satellite of the Saturne planet) of the "atomic beam" type producing either protons or deuterons with either vector or tensor polarization. A heavy-ion source (CREYBIS) for production of ions up to mass 40 is now being tested.

The first successful acceleration of polarized protons at SATURNE have recently been achieved. Below is a preliminary account of these tests but one must be aware that numerous items are not yet fully optimized so that the final results will hopefully be more promising.

A. POLARIZED SOURCE

The main elements are as follows:

1. - Dissociator

It dissociates the molecular gas H_2 into hydrogen atoms H. It is made of a U-shaped Pyrex tube with a 2.5 mm diameter nozzle. The dissociation is induced by a 20 MHz oscillator of 6 Kw (peak power). Both the oscillator and the gas injection are pulsed (duration 30 ms) at the synchrotron acceleration frequency of one burst every second at energies below 1 GeV to one burst every 3 seconds at 3 GeV. The atomic beam size is reduced by a set of 3 skimmers of diameters 2.2, 3.5 and 6.5 mm before entering the sextupole magnet.

2. - Sextupole magnet

It selects and focuses the atoms in those states with a positive effective electronic magnetic moment, $|m_e, m_p\rangle = |1/2, \pm 1/2\rangle$, where m_e and m_p are respectively the spin projections of electrons and protons as shown in Fig. 1. The sextupole is a one-piece magnet 50cm long with entrance and exit diameters of 7 and 14 mm respectively. The maximum field at the pole tips is 9000 oe.

---

*In collaboration with: R. Beurtey, J. Bony, J. Bourbonneux, P.Y. Beauvais, R. Burgei, P.A. Chamouard, M. Grand, R. Maillard and R. Vienet for the atomic source; G. Milleret and J.M. Soudan for the polarimeter; E. Grorud, J.L. Laclare, G. Leleux, A. Nakach and A. Ropert for the tests of resonance crossing.

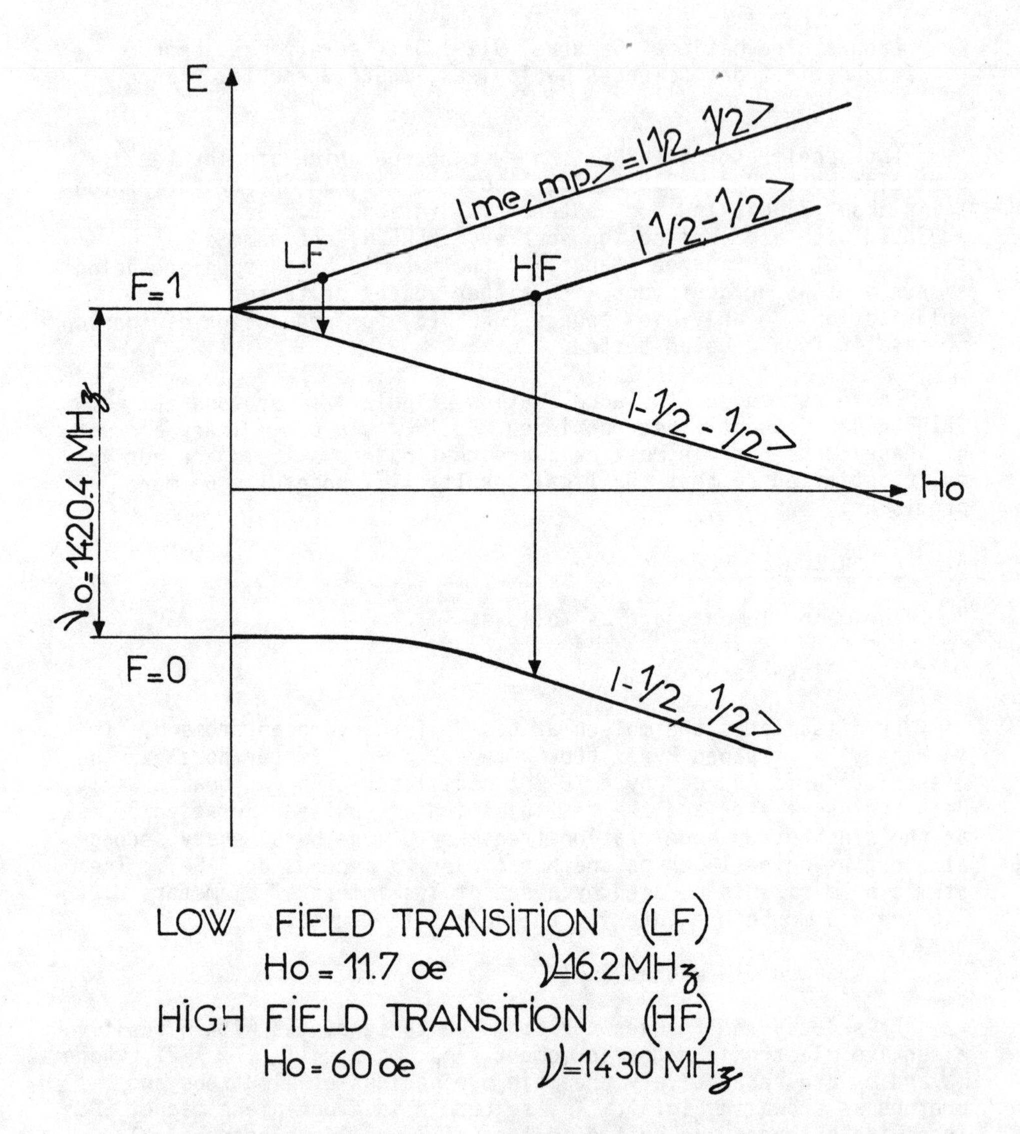

Fig. 1. Energy of atomic states as a function of static magnetic field H_0 and characteristics of RF transitions (not to scale).

3. - Radio-Frequency (RF) transition units

Their purpose is to induce transitions between magnetic sub-levels in order to produce fully polarized protons of either sign. There are two types of transitions.

A "low-field" RF transition unit induces the transition $|1/2, 1/2\rangle \rightarrow |1/2, -1/2\rangle$. The unit is a coil of diameter 30 mm and length 48 mm oscillating at 16.2 MHz, through which passes the atomic beam. The resulting RF field H_1 is perpendicular to a static field H_0 = 11.7 oe. The variation of the static field across the coil which ensures a fully adiabatic transition is $\Delta h = \pm 3$ oe. The degree of polarization is optimum for the RF power in the range of 1 to 4 watts. After ionization in a strong magnetic field, the resulting protons have their spins oriented opposite to the magnetic field direction ($m_p = -1/2$).

The "high field" RF transition unit induces the transition $|1/2, -1/2\rangle \rightarrow |1/2, 1/2\rangle$. This is done with a coaxial cavity operating at 1430 MHz in a static field H_0 = 60 oe parallel to H_1 ($\Delta h = \pm 5$ oe). The optimum power is about 10 watts. The resulting proton beam has spins oriented in the direction of the magnetic field.

The static field H_0 and the gradient Δh necessary for both transitions are produced with a single C-shaped permanent magnet made of soft iron pole pieces and ferrite rings.

The transitions can be flipped every one or two bursts by a clock synchronized with the synchrotron timing and commencing 20 ms before the acceleration cycle.

The atomic beam intensity at the exit of the ionizer measured with a compression gauge is about $(2 \pm 1) \times 10^{16}$ atoms-s^{-1}.

4. - Ionizer

It has been delivered by ANAC Inc. (Santa Clara, CA 95051, USA) {1}. Ionization is produced by a reflex electron beam in a longitudinal field produced by 6 independently adjustable coils.

Some initial difficulties are probably due to the fact that the ionizer was used in a horizontal position while the customary mounting is in a vertical position. It has now been modified for better mechanical stability and correct magnetic alignment has beem obtained by firm mechanical tightening of the coils and careful adjustments of the high-voltage electrodes.

The ion beam is extracted at a potential of 13 KV. Intensities obtained so far are of the order of 45 μA of total ionic current.

At the exit of the ionizer the proton beam is electrically deflected through 90° by a parabolic mirror. The spin vector lies then in the horizontal plane. it is aligned upwards or downwards (depending on the RF transition in use) in a 17 cm long solenoid having a static field of 500 oe at its center.

The whole source assembly is held in a high-voltage terminal at an HV of 170 KV for protons. After extraction at ground the beam is deflected into an injection line leading to the Linac operating in the $2\beta\lambda$ mode where it is accelerated up to 5 MeV per nucleon and finally injected into the synchrotron. Certain diagnostic elements such as variable slits, a current meter and an emitance meter allow some optimization of the beam phase-space in order to match the Linac acceptance. Intensities of up to 35 μA of real polarized beam (difference between sextupole ON and OFF) have been obtained after the electrostatic mirror at 13 KV, and 9μA has been accelerated in the Linac. These currents may be improved in several ways. Without modifying the hardware, a better understanding of the atomic source and ionizer should lead to optimum tuning since the ionizer operates through discrete jumps which require long set-up perionds {1}. Improvements to the injection line are also expected in the near future. Some improvements in the polarized source itself are also being studied: a new dissociator (eventually He-cooled), a new sextupole divided into independent sections. The intensity on target is (4-5) $\times 10^8$ p per burst. Some losses are due to the Linac efficiency (66%), synchrotron injection efficiency (20-50%), capture efficiency (25-40%), and extraction efficiency (50-80%) giving an overall efficiency from Linac to target of 1-10% depending on tuning. The main loss is due to the fact that Saturne is a pulsed machine which accelerates particles injected during ~ 300 μs every burst, resulting in an efficiency of a few 10^{-4} which cannot be raised without some sort of storage. A pre-injection storage ring (project MIMAS) should improve intensities for polarized beams (and also for heavy-ions) by a factor of 20 to 100.

B. POLARIMETER

A polarimeter has been installed in an independent beam line at extraction point SD2. It uses well-known proton-proton elastic scattering polarization data {2-6}. A movable 4-position target ladder contains: a 2 cm-wide CH_2 target, a 2 cm-wide C target, and a 4 mm-wide CH_2 strip target. The fourth position has no target at all for background measurements. Target thicknesses range from 100 to 400 mg-cm^{-2} depending on energy.

The detector system is made up of 4 arms as shown in Fig. 2. Scattered protons are detected at symmetric forward angles by a 3-scintillator counter telescope. Conjugate recoil protons are detected at backward angles by a 2-counter telescope. Each scintillator is 34 mm wide and 84 mm high giving an angular aperture of ± 1° in the horizontal plane and ± 2.5° in the vertical plane; the solid angle is 3×10^{-3} sr for the forward defining counters. Detectors F1, F2 and B1 (F=Forward, B=Backward) are made of 4 mm thick NE 104 scintillating material; the last detector within each arm, F3

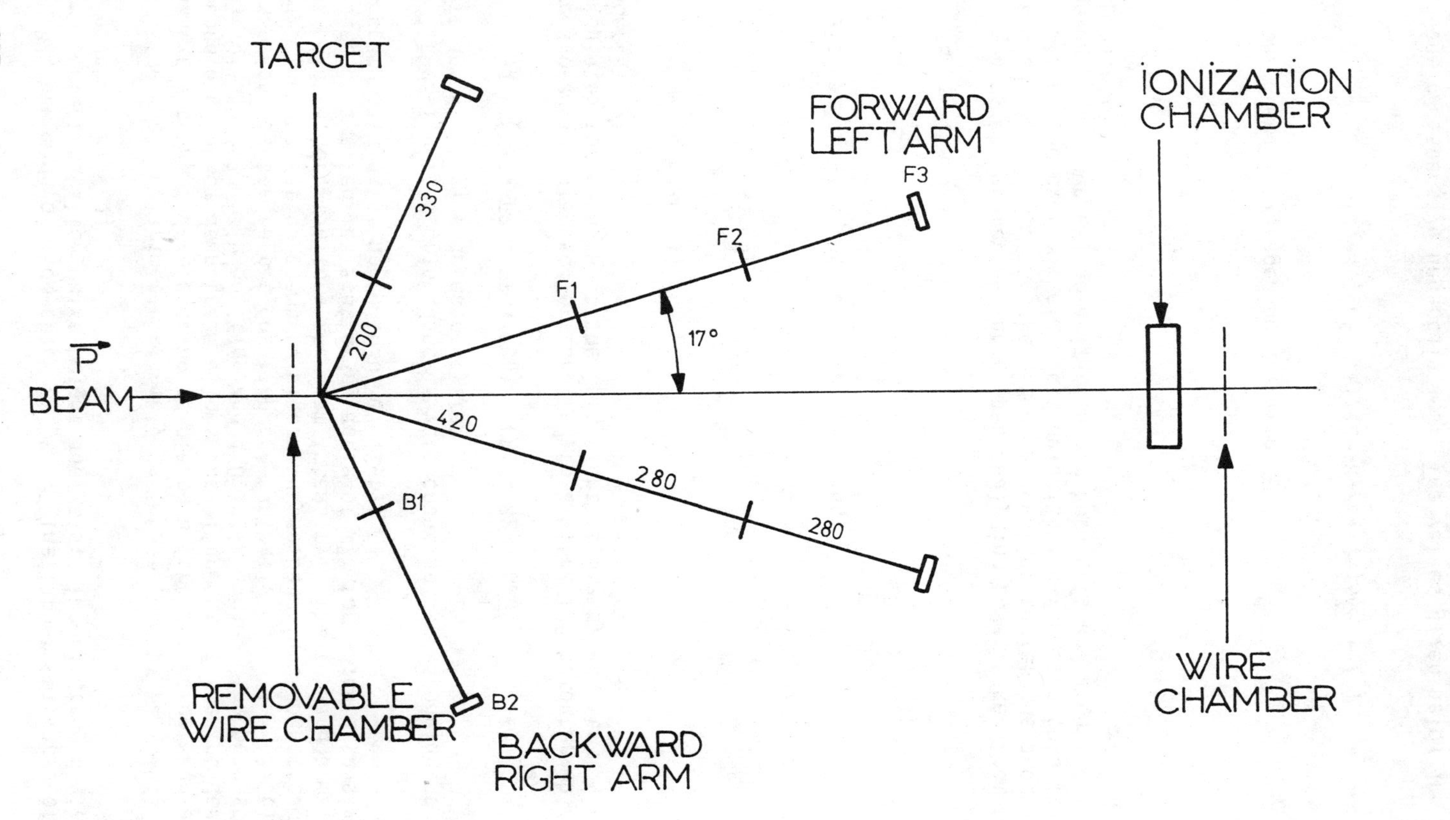

Fig. 2. Diagram of the polarimeter assembly. Dimensions are indicated in cm.

and B2, are made of 1 cm-thick Pilot U. The larger thickness of these detectors allow pulse-height analysis. The electronic diagram is given in Fig. 3.

The whole polarimeter assembly is in air. An XY removable wire chamber with 2 mm wire spacing is placed before the target and a second one 150 cm downstream allows the alignment of the incident beam within ± 1 mm in position and ± 0.1° in angle. The beam spot dimensions at the target position are typically (X x Y) = (3× 6) mm^2 FWHM. An ionization chamber, placed in the beam just downstream of the polarimeter, monitors the beam intensity. The forward arms are placed at a fixed lab scattering angle of 17° for which extensive and accurate data exist up to 1030 MeV {2-5}.

The five-fold coincidence + veto counting corresponding to the Left and Right side detector system

$$L,R = (F1 \cdot F2 \cdot F3 \cdot F3_V) \times (B1 \cdot B2 \cdot \overline{B2_V})$$

are stored in Camac scalers and fed to a Mitra 125 computer which calculates for each burst the left-right asymmetry and its uncertainty

$$A_{LR} = (L-R)/(L+R)$$

$$\Delta A_{LR} = \{(1-A_{LR}^2)/(L+R)\}^{1/2}.$$

It also calculates the polarization using

$$P_{LR} = A_{LR}/A$$

$$\Delta P_{LR} = \Delta A_{LR}/A$$

where A is the energy-dependent p-p analyzing power taken from the measured data {2-5}. At the end of a counting series the program calculates the overall Left-Right polarization for each spin state, the up (↑) - down (↓) polarization for both Left and Right separately

$$P = \frac{1}{A} (\uparrow-\downarrow)/(\uparrow+\downarrow)$$

and the overall averaged polarization

$$P = \frac{1}{A} \{(L\uparrow \cdot R\downarrow/L\downarrow \cdot R\uparrow)^{1/2} - 1\}/\{(L\uparrow \cdot R\downarrow/L\downarrow \cdot R\uparrow)^{1/2} + 1\}$$

and their respective uncertainties. The values given below are for a

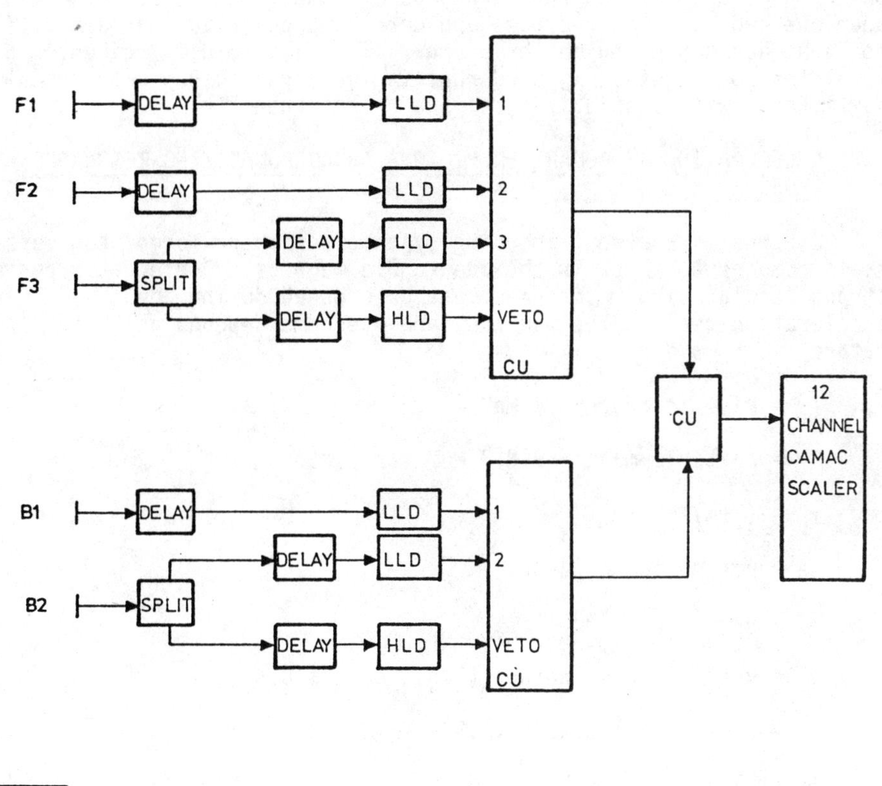

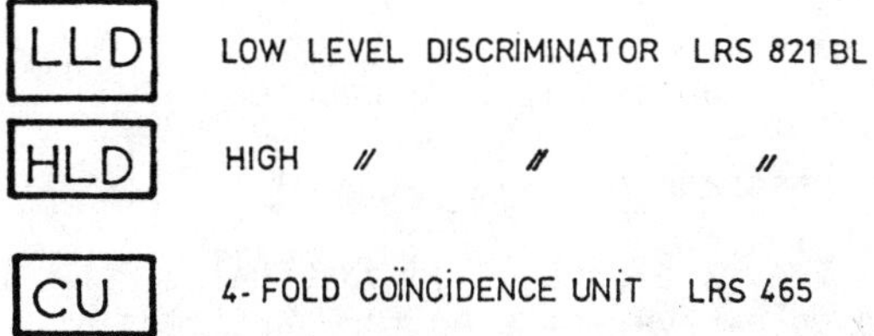

Fig. 3. Schematic diagram of the polarimeter electronics for conjugate arms.

CH_2 target, uncorrected for carbon background. The effect of C has been checked at a few energies and corrected polarizations are 0.1% to 3% higher depending on the energy. The five-fold coincidence level for no-target background runs is less than 10^{-3}. Geometrical asymmetries with unpolarized beams are less than 1%.

C. ACCELERATION OF POLARIZED PROTONS AND DEPOLARIZING RESONANCE CROSSING

Saturne is a strong-focusing synchrotron where large transverse field components exist in the quadrupole magnets. Different types of strong depolarizing resonances must be crossed during the acceleration cycle. The way they are overcome depends on their nature. Let us define

T = kinetic energy in MeV

M = particle mass = 938.3 MeV for protons

$\gamma = (T+M)/M$

g_p = proton Lande factor

$G = g_p/2-1 = 1.7935$

ν_z = vertical wave number ($3.5 < \nu_z < 3.8$)

ν_x = radial wave number ($2.5 < \nu_x < 3.8$)

Central value $\nu_z = \nu_x = 3.66$

The following depolarizing resonances may be encountered:

1. First order systematic resonances.

They are strong and wide. They are crossed adiabatically; the problem is to ensure a complete spin-reversal even for particles at the center of the beam-core {7}.

a) Closed orbit resonances. They occur when $\gamma G = n$. The list for n=2-7 is given in Table 1 (see next page) with the corresponding static widths calculated for a closed orbit defect of 4 mm and a depolarization of 1%.

The cases n=2 and n=3 have been studied in detail. For n=2 the natural closed orbit has small vertical amplitudes ($z_{c.o} < 2$ mm) leading to a partial depolarization. The second harmonics are increased by strongly bumping the beam vertically between 100 and 120 MeV when its energy approaches the resonance energy. This is done with a set of 12 closed orbit correction windings, located in the defocusing quadrupoles, fed by 12 D.C. power supplies pulsed rapidly and simultaneously within 2 ms by the same command signal {9}.

TABLE 1

Energy E and static widths ΔE (in MeV) for first-order betatron resonances ($\gamma G = \nu_z$ and $\gamma G = 8 - \nu_z$, $\nu_z = 3.66$) and closed orbit resonances ($\gamma G = n$) up to 3 GeV. ΔE is computed for a 1% depolarization effect.

γG	E	ΔE
2	108	± 10
3	631	± 14
ν_z	977	± 50
4	1155	± 18
$8-\nu_z$	1333	± 34
5	1679	± 21
6	2201	± 25
7	2725	± 28

Displacements of 2.7 mm per amp are obtained and the maximum current is ± 8 amps. Fig. 4 shows the polarization at 796 MeV as a function of the current for 3 different phases of the second harmonics closed orbit which is shown in Fig. 5. For I=0 amps one observes a partial depolarization without spin-reversal. For some phases a complete correction without spin-reversal can be obtained but it is a very sensitive function of the closed orbit correction. For large vertical deformations a complete spin-reversal can be obtained in all 3 cases. A partial depolarization, which is not yet fully understood, seems to occur for some intermediate values of the current. In Fig. 6 we compare the beam polarizations at 520 and 796 MeV for phase I. One observes a change of the polarization sign due to the third harmonics at 630 MeV; the natural width is large enough to ensure a complete spin-reversal without any closed-orbit third harmonics correction.

b) <u>Betatron resonances.</u> They appear for $\gamma G = \nu_z$ around 1 GeV and $\gamma G =$ <u>$8-\nu_z$ around 1300 MeV</u>. They have large static widths ΔE, respectively about ± 50 MeV and ± 34 MeV where ΔE is computed for a 1% depolarization and $\Delta z = 10$ mm. They have not yet been fully studied.

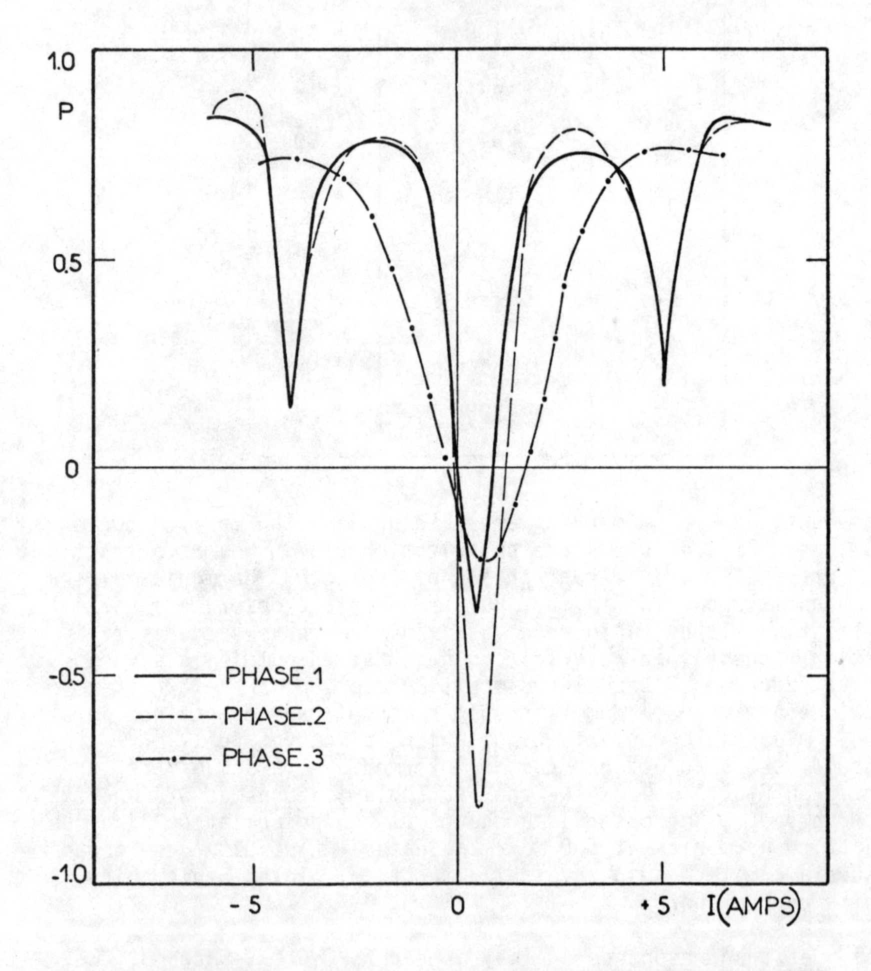

Fig. 4. Beam polarization at 796 MeV as a function of the current for 3 different phases of the second harmonics closed orbit. The corresponding vertical bumping is $\Delta z_{c.o} = 2.7$ mm per amps.

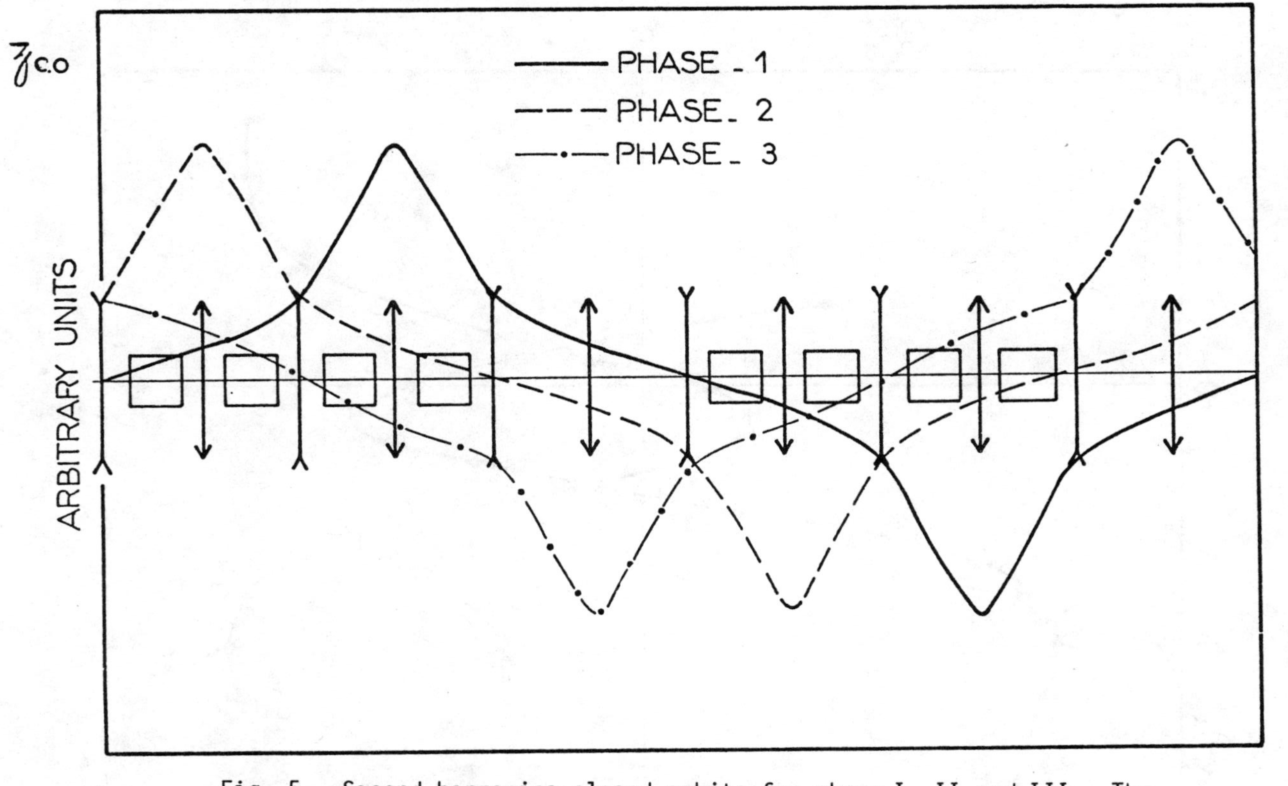

Fig. 5. Second harmonics closed orbits for phase I, II and III. The horizontal axis corresponds to a half-machine cell. Phases are shifted by 60°. Dipole magnets are represented by square boxes and vertically focusing (defocusing) quadrupoles by closed (open) arrows.

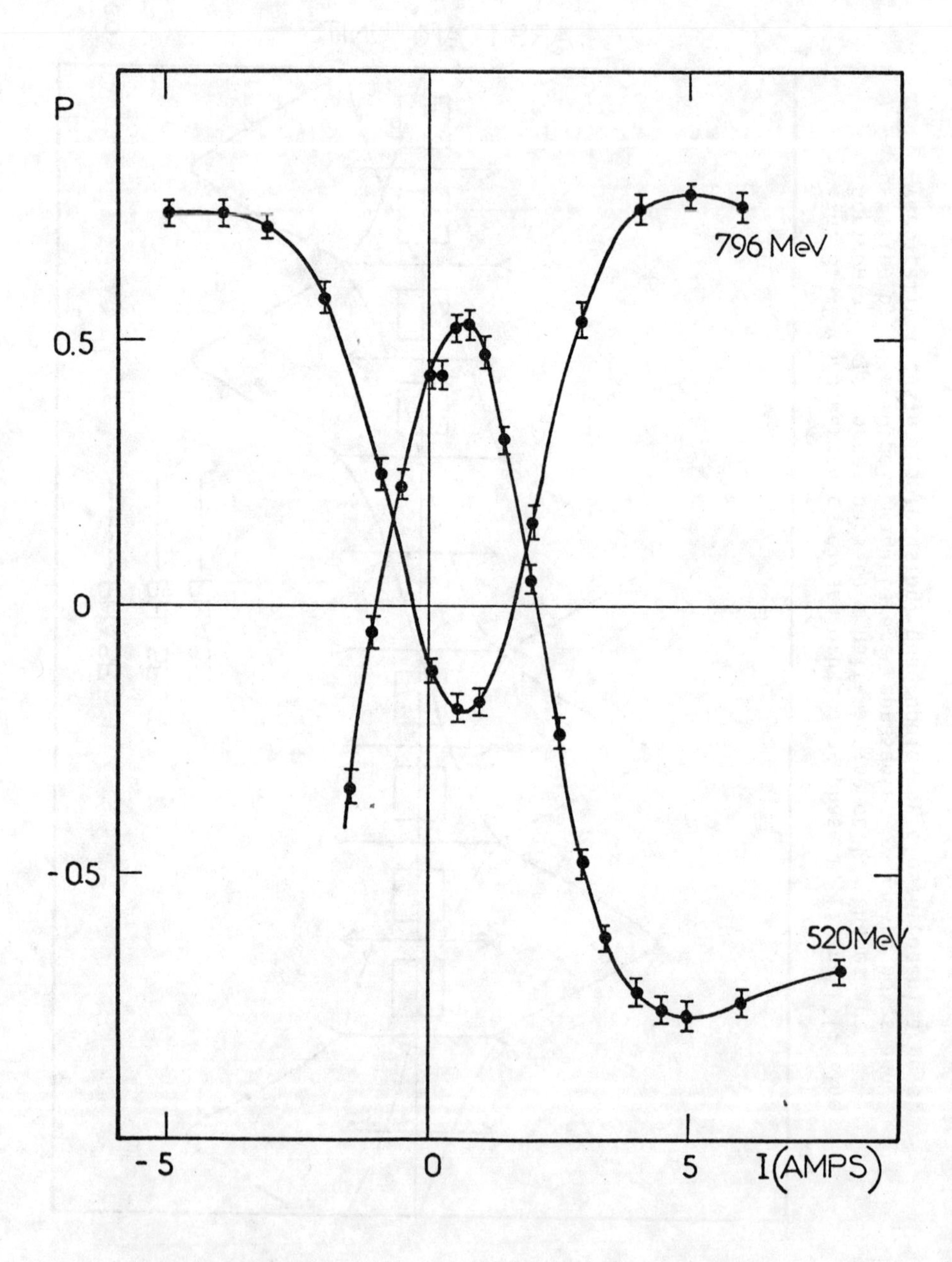

Fig. 6. Beam polarization at 520 and 796 MeV for a phase I second harmonics closed orbit.

A summary of all first order systematic resonance energies and widths is given in Table 1.

2. Linear non systematic resonances

They are due to the non-identity of the quadrupole magnets. They are narrow and can be easily overcome by choosing the appropriate acceleration tuning. We have studied in detail the resonance $\gamma G = 7 - \nu_z$. In Fig. 7 we give the polarization for 3 different energies, namely 775, 825 and 860 MeV as a function of ν_z at flat-top. By choosing a correct ν_z value, depolarizing resonances can be completely avoided and a maximum polarization obtained at any energy.

3. Second order resonances

They are due to non-linearities in dipole and quadrupole magnets and they primarily affect the outer region of the beam for large betatron amplitudes {8}.

a) Hexapolar effects in dipoles

The resonance

$$\gamma G = \pm \nu_z \pm \nu_x - 4n$$

occurs at 800 MeV for n=1 and at 1500 MeV for n=3.

b) Non-linear stray-field in quadrupoles

Some of these resonances are:

second order: $\gamma G = \pm \nu_z \pm \nu_x - 4n$

third order: $\gamma G = \pm 3\nu_z - 4n$
$\gamma G = \pm 2\nu_z \pm \nu_x - 4n$

fourth order: $\gamma G = \pm \nu_x \pm 3\nu_z - 4n$
$\gamma G = \pm \nu_z \pm 3\nu_x - 4n$

In the Saturne Synchrotron non-linear effects have been carefully corrected so that all quadrupole field gradient lengths are within 5×10^{-4} and thus the second order resonances are weak and narrow. Calculations predict depolarization effects of 10^{-3} to 10^{-4}. They have not given any trouble up to 1 GeV.

D. CONCLUSIONS

Polarized protons up to 8×10^8 particles per burst have been successfully accelerated in the Saturne synchrotron at energies up to 1060 MeV. Crossing of first- and second-order resonances has been

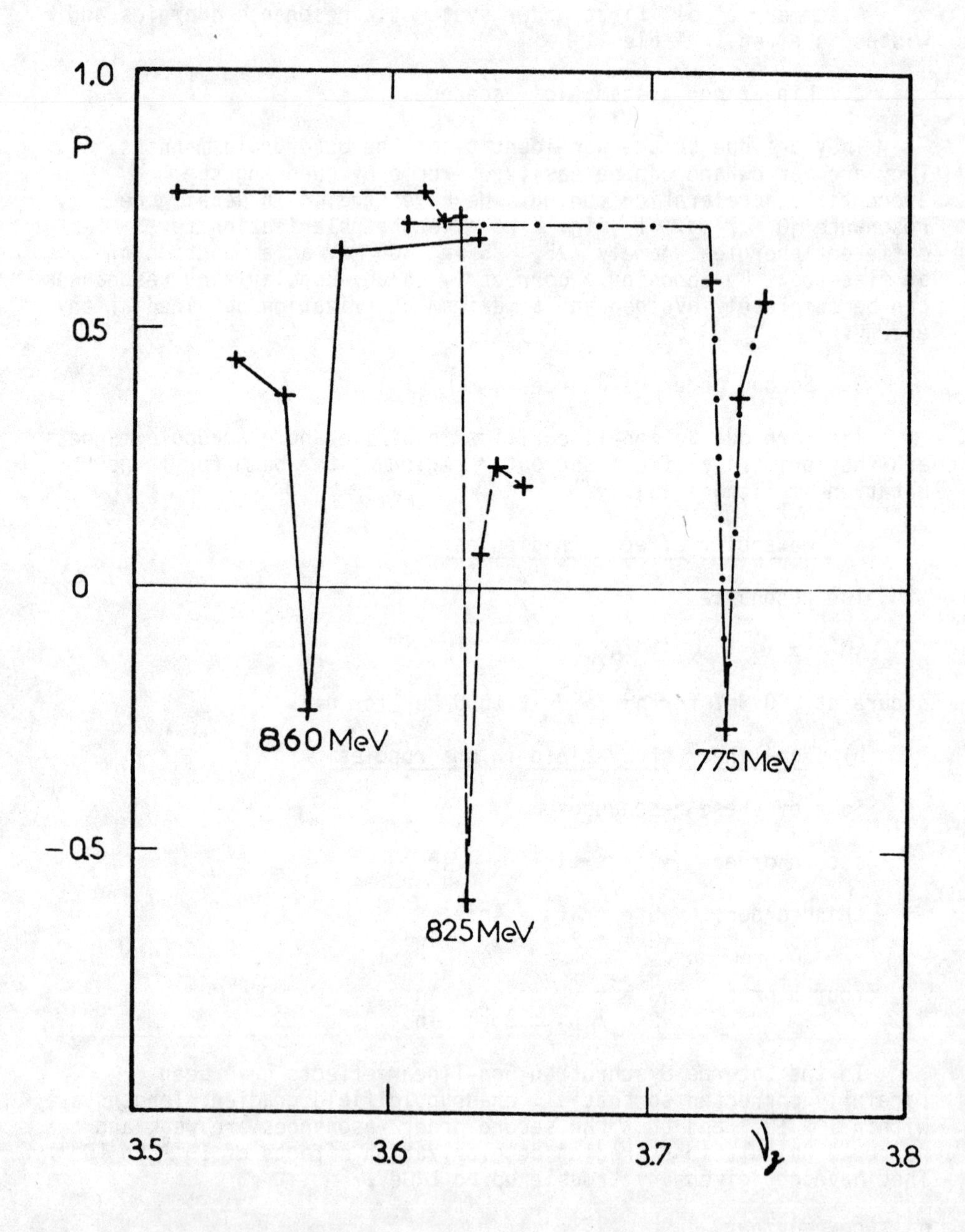

Fig. 7. Beam polarization at 775, 825 and 860 MeV as a function of ν_z at flat-top. The effect of the non-systematic linear resonance $\gamma G = 7-\nu_z$ is to induce a strong depolarization and partial spin flip.

achieved and beam polarizations close to 80% have been measured up to 900 MeV; but some depolarization arises at higher energy. The problem is under scrutiny and improvements in both the intensity and the polarization at energies above 1 GeV may be expected in the near future.

ACKNOWLEDGMENTS

This project would have not been achieved without the effort of numerous technicians and engineers of the Laboratoire National Saturne. We thank in particular the "Service Machine" for succeeding in accelerating very low intensity beams.

REFERENCES

1. H.H. Glavish, IEEE Transactions Nucl. Sciece. NS-26, 1, (Feb. 1979).
2. L.G. Greeniaus, D.A. Hutcheon, C.A. Miller, G.A. Moss, G. Roy, R. Dubois, C. Amsler, D.K.S. Koene and B.T. Murdoch, Nucl. Phys. A322, 308 (1979).
3. P.R. Bevington, M.W. McNaughton, H.B. Willard, H.W. Baer, E. Winkelmann, F. Cverna, E.B. Chamberlin, N.S.P. King, R.R. Stevens, H. Wilmes and M.A. Schardt, Phys. Rev. Lett 41, 384 (1978).
4. N.W. McNaughton, P.R. Bevington, H.B. Willard, E. Winkelmann, E.P. Chamberlin, F.H. Cverna, N.S.P. King and H. Willmes, Phys. Rev. C23, 1128 (1981).
5. M.L. Marshak, E.A. Peterson, K. Ruddick, J. Lesikar, T. Mulera, J. Roberts, R. Klem, R. Talaga and A. Wriekat, Phys. Rev. C18, 331 (1978).
6. H.E. Neal and M.J. Longo, Phys. Rev. 161, 1374 (1967).
7. E. Grorud, J.L. Laclare and G. Leleux, Rapports LNS-GERMA 75-47 and 75-48, Saclay (1975), (unpublished).
8. E. Grorud, J.L. Laclare and G. Leleux, Rapport LNS 20, Saclay (1978), (unpublished).
9. L. Degueurce and A. Nakach, 1981 Particle Accelerator Conference, March 11-13, 1981, Washington DC, (to be published).

Note added in proof: The first tests of the polarized proton beam described in this paper have been done with B=2.1 Tesla·s^{-1} which is half the nominal value. Further tests with a nominal value of 4.2 Tesla·s^{-1} have demonstrated that the $\gamma G=2$ (at 108 MeV) and $\gamma G=3$ (at 631 MeV) closed orbit resonances can be fully corrected without spin-flip and polarizations up to + 89% have been obtained at 725 MeV.

A strong $\gamma G=7-\nu_z$ resonance due to a quadrupole defect enhancement for a nominal tune near $2\nu_z=7$ has been observed between 700 and 900 MeV, leading to a 10-15% depolarization at 900

MeV. A complete correction (within 1-2%) was obtained using auxiliary quadrupole magnet circuits and polarizations up to +87% have been measured at 900 MeV without any spin-flip.

Finally the $\gamma G = \nu_z$ intrinsic resonance (around 977 MeV) has been crossed with full spin-flip for both B=2.1 and 4.2 Tesla$\cdot s^{-1}$. In conclusion, polarizations up to 81% have been obtained at 1001 MeV indicating a depolarization of less than a few per cent over the whole energy range. In the process, the beam had to pass through two fully corrected closed-orbit resonances (γG=2 at 108 MeV and γG = 3 at 631 MeV), one partially corrected linear non systematic resonance ($\gamma G=7-\nu_z$ around 850 MeV) and a linear intrinsic resonance ($\gamma G=\nu_z$ at 977 MeV) crossed with spin-flip.

KEK POLARIZED BEAM PROJECT AND POLARIZED ION SOURCE

Y. Mori, K. Ito, A. Takagi and S. Fukumoto
National Laboratory for High Energy Physics
Oho-machi, Tsukuba-gun, Ibaraki-ken, 305, Japan

INTRODUCTION

There has been some interest in the acceleration of polarized protons in the KEK 12 GeV synchrotron from the beginning of its construction.[1] The KEK 12 GeV synchrotron consists of four accelerator stages: the 750 keV Cockcroft-Walton preinjector, the 20 MeV linac, the 500 MeV booster synchrotron and the 12 GeV main ring.

In order to accelerate polarized protons in this machine, it is first necessary to overcome a very strong depolarizing resonance at the energy of 239.3 MeV (γ_{res}= 1.2552) in the booster synchrotron. Previously, it was thought that this resonance was so strong that it was impossible to accelerate polarized protons in the KEK machine.[2] Recently, however, this depolarization resonance was re-examined using a new concept of spin-flip and then it was found that the proton spin could be completely flipped without destroying the polarization, as shown in Fig. 1. Thus, the acceleration of polarized protons in the KEK machine might be possible. The strength of the depolarizing resonances in the main ring has also been calculated rigorously. In this way, the project of acceleration of polarized protons in KEK was started last year and a new 750 keV preinjector for the polarized ion source is presently under construction.

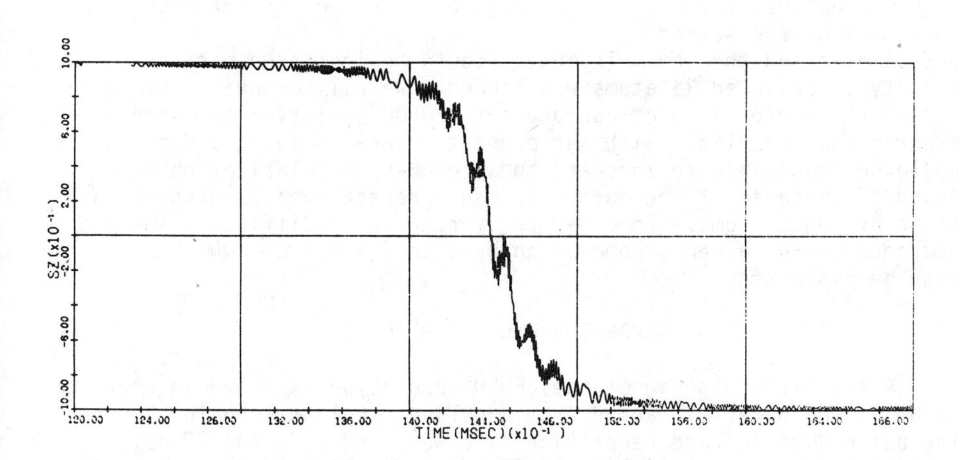

Figure 1. Calculation of polarization in the booster synchrotron using spin-flip at the resonance.

It is well known that for a proton synchrotron, H^- ions are more advantageous than H^+ ions because of the possible use of the charge exchange multi-turn injection technique. In KEK, an H^- injection scheme at the entrance of the booster is now being studied and some calculations show that about 200 turns could be accepted and that about half of the beam could be accelerated up to 500 MeV. The rest is lost by multiple scattering in a carbon foil. This means that an H^- ion beam is apparently equivalent to an about 20 times more intense H^+ ion beam. Thus we have concentrated on developing a polarized H^- ion source. Previously, we developed a Lamb-shift type polarized H^- ion source; however, because the space charge fields of charged particles quenched the metastable atoms, it seemed impossible to get a beam intensity of more than 1 μA.[3]

Last year, we started to develop a new type of polarized ion source which utilizes the charge-exchange reactions between a fast proton beam and electron-spin oriented Na atoms.[4] The principle of this polarized ion source is shown in Fig. 2. We named this polarized ion source APOLON (Advanced POLarized ion source with Oriented Na atoms). There are two ways of producing electron-spin oriented Na atoms: an inhomogeneous magnetic field scheme with multi-pole magnets[5] and an optical pumping scheme using a dye laser beam.[6] We have tested each scheme and found that the latter was superior for several reasons: the structure of the source was simpler and the beam intensity could be increased because the density of oriented Na atoms was higher. We will describe in some detail the performance of our present APOLON prototype and some experimental results. With our present apparatus we consider that it would be impossible to increase the proton-spin polarization above 40 to 50% because of the quasi-resonant character of forming the n=2 state hydrogen atoms. In order to overcome this difficulty, we are now considering a new scheme which uses an E × B field which will also be discussed.

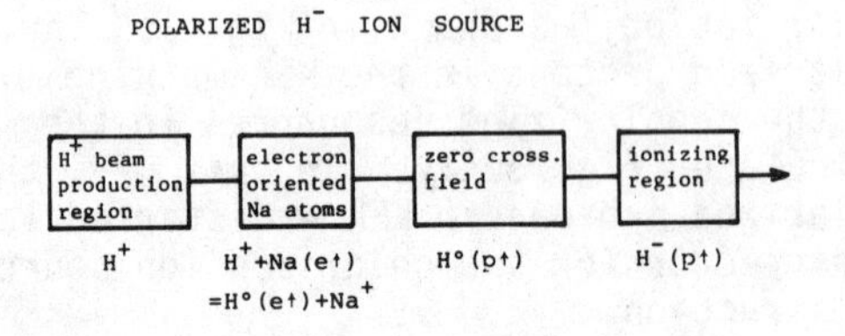

Figure 2. Principle of APOLON

APPARATUS AND EXPERIMENTS

A schematic diagram of the APOLON prototype is shown in Fig. 3. An H^+ ion beam was extracted from a pulsed duoplasmatron ion source. The pulse duration and repetition rate were 150 μsec and 20 Hz, respectively. A potential of 20-25 kV was applied to the extraction electrode.

The optical pumping region contains a Na cell and an oven. The cell was made of copper and placed inside the solenoid which produced a longitudinal magnetic field of 5 kG maximum. A sheath-like heater

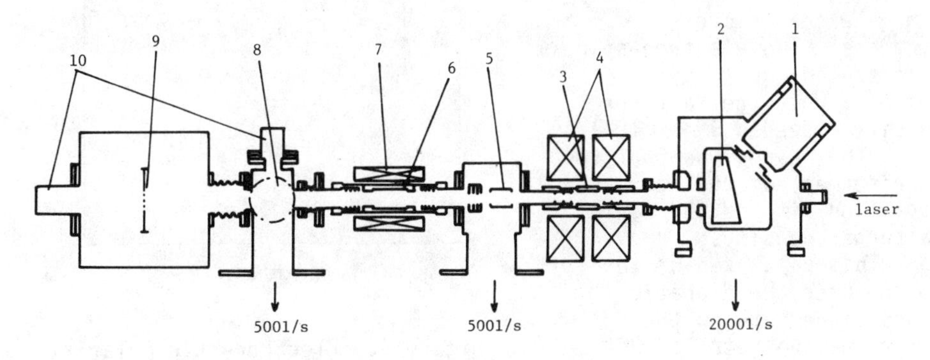

1. duo plasmatron
2. 45° bend. magnet
3. Na optical pumping cell
4. solenoid coil
5. deflection plates
6. Na ionizing cell
7. solenoid coil
8. 90° analysing magnet
9. profile monitor
10. Faraday cup

Figure 3. Schematic layout of APOLON

wound around the cell was heated to about 400°C to prevent the Na atoms from depositing on the cell walls. Freon-cooled traps placed at the entrance and the exit of the cell prevented the Na atoms from escaping.

Electron-spin oriented Na atoms were produced by means of a dye laser with R6G tuned to the Na D1 line. The dye laser operated with the line width of the axial modes of about 30-40 GHz and an output power of 1W. The electron-spin polarization was measured using a 6-pole magnet. A small fraction of oriented Na atoms streamed through a 6-pole magnet and was then detected by a surface ionization detector. The current of the detector, I, varied when a quarter wave plate was rotated which changed the vertically polarized light emitted by the dye laser to a circularly polarized one. The electron-spin polarization was then estimated by the following equation:

$$P = \frac{\varepsilon}{A}$$

where $\varepsilon = [I(max)-I(min)]/[I(max)+I(min)]$ and A is the analyzing power of 6-pole magnet. I(max) and I(min) are the maximum and minimum ion currents detected by the surface ionization detector. The analyzing power of the 6-pole magnet, A, was estimated to be 0.47 using a Monte Carlo simulation program which took into account the dependence of the effective magnetic moment on the magnetic field. Fig. 4 shows the behavior of the electron spin polarization of Na atoms as a function of target density in the cell. The target density was calculated from the H^- ion current using the $H^+ \rightarrow H^\circ$ and $H^\circ \rightarrow H^-$ charge-exchange cross sections. The electron-spin polarization decreased abruptly when the Na target density exceeded $3\text{-}4 \times 10^{11}/cm^3$. The reason was that the imprisonment of the resonance radiation limited the useful Na target density.

Figure 4. Electron-spin polarization of Na atoms as a function of the target density.

This polarized H^- ion source used the diabatic transitions between the hyperfine substates of H° atoms to transfer the electron-spin polarization to the proton-spin polarization. In this scheme, the proton-spin depolarization depends on the gradient of the magnetic field near the zero-crossing point. This effect was known as a Majorana depolarization, and was estimated by Ohlsen[7] for Lamb-shift type sources. We estimated these depolarizations for our scheme. Fig. 5 shows the calculated values of the proton-spin polarization of 5 keV H° atoms initially in the hyperfine substate of $m_J = +1/2$ and $m_I = +1/2$ moving at a distance of 1 cm from the beam center. It was found that the magnetic field gradient near the zero-crossing point should be less than 1.0 G/cm. Fig. 6 (see next page) shows the dependence of the proton-spin polarization on the magnetic field decay length (from strong field to weak field). It showed that the decay length should be more than 2 cm for keeping the beam polarized.

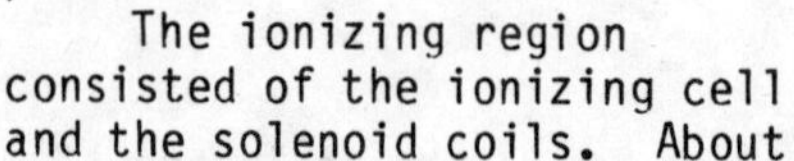

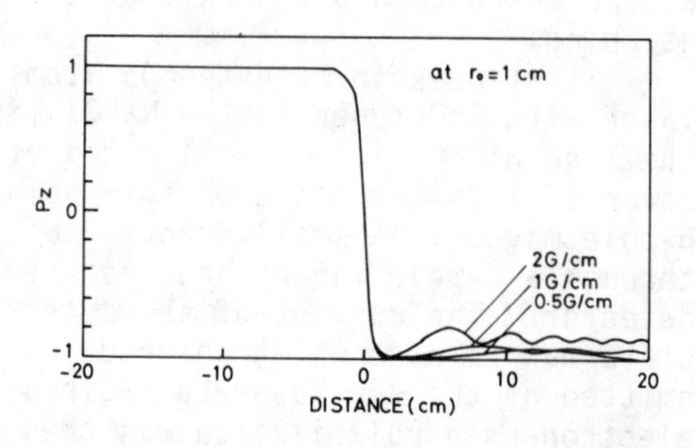

Figure 5. Diabatic transitions in zero-crossing magnetic field.

The ionizing region consisted of the ionizing cell and the solenoid coils. About 80% of H° atoms were converted to H^- ions. We have obtained polarized H^- ion beams of 4 to 5 μA. The fraction of background unpolarized H^- ions generated by the charge-exchange reactions with residual gases was about 10%.

POLARIZATION ESTIMATION

The polarized H^- ion source utilized the charge-exchange reaction between a fast H^+ ion beam and optically pumped Na atoms. This charge-exchange reaction showed a quasi-resonant character for the process of forming the n = 2 states of H° atoms in this energy region. The n = 2 states of H° atoms are the 2S and 2P states. If a polarized electron is captured in the 2P state, then the electron-spin polarization is destroyed by a spin-orbit coupling force. This depolarization effect (which varied as a function of magnetic field strength) was calculated using the wave functions of the 2P substates. The decreased polarization was obtained from the following equations:

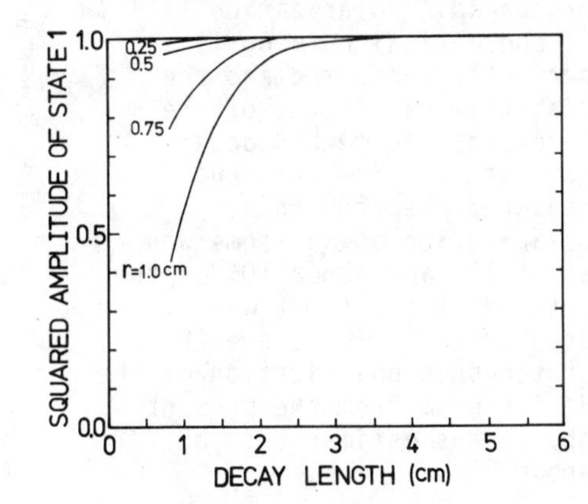

Figure 6. Dependence of the proton-spin polarization on the magnetic field decay length.

$$P = (1 + \delta_1^2 + \delta_2^2)/3 \text{ , where}$$

$$\delta_1 = (\xi - \frac{1}{3})/\sqrt{1 + \frac{2}{3}\xi + \xi^2}$$

$$\delta_2 = (\xi - \frac{1}{3})/\sqrt{1 - \frac{2}{3}\xi + \xi^2}$$

where $\xi = \mu_0 B/\Delta E$. ΔE is the fine structure energy splitting between $2P_{3/2}$ and $2P_{1/2}$ states. The broken curve in Fig. 7 showed the values calculated from the above equation.

Fig. 8 (see next page) shows the behavior of the extracted polarized H^- ion current as a function of the magnetic field strength. The beam current increased at first and then above 1 kG it

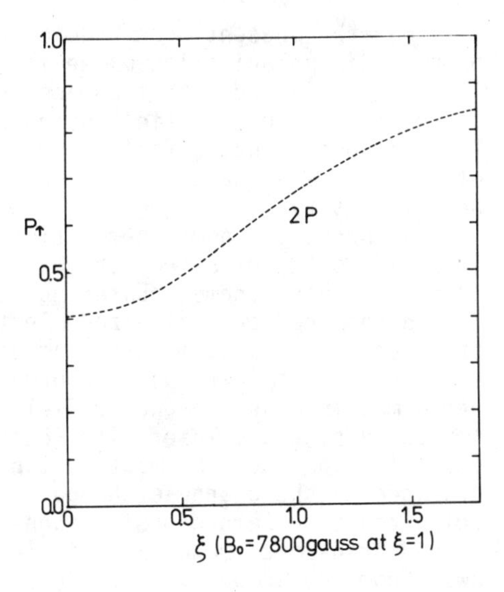

Figure 7. Plot of the electron-spin polarization of H° atom as a function of the magnetic field strength.

decreased gradually because of the emittance blow-up. When the magnetic field strength is about 2 kG, the ideal proton-spin polarization is 45% if the optical pumping is perfectly performed and the diabatic transitions of the zero-crossing method occur completely. However, the measured electron-spin polarization of Na atoms was about 75% and about 10% of the total H^- ion current was unpolarized. Therefore the proton-spin polarization of the H^- ion beam from the present source was estimated to be about 30%.

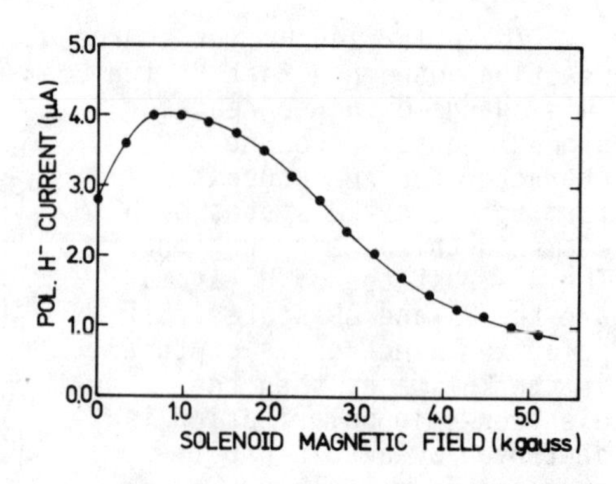

Figure 8. Plot of the extracted polarized H^- ion current as a function of the magnetic field strength.

A NEW SCHEME USING AN E × B FIELD CONFIGURATION

For the present scheme described above we cannot expect to get a proton-spin polarization exceeding 40-50% without losing beam intensity. In order to increase the polarization, it is necessary to increase the magnetic field strength at the charge-exchange region. The solenoid magnetic field, however, produces an emittance blow-up of the charge-exchanged hydrogen beam which abruptly decreases the beam intensity.

We propose a new scheme using an E × B configuration which, ideally, would not affect the beam optics. Fig. 9 shows a schematic layout of this scheme. Entering into the E × B field region, a 5 keV H^+ ion captures the polarized-electron from a Na atom flowing out of the oven. The Na atoms are exposed transversely to a dye laser beam in order to eliminate Doppler broadening, and optical pumping is performed in a weak magnetic field of about 100 G. A single frequency ring dye laser with feedback loop for frequency stabilization would be most suitable for this scheme. H° atoms produced by the charge-exchange reactions can be proton-spin polarized by a zero-crossing magnetic field which is generated by a second transverse magnet. This magnet also sweeps charged particles away from the H° beam. The field gradients of the magnetic fields must be appropriately adjusted to maintain the adiabatic condition by using two soft iron plates. The ionizing region occurs in the solenoid magnetic field of 2 kG. The direction of the magnetic field should be changed from transverse to longitudinal to maintain the adiabatic condition in the strong magnetic field.

In this scheme, in principle, beam optics are almost unaffected even if the magnetic field is high. But the energy of the beam would be spread. This energy spread can be estimated by solving the

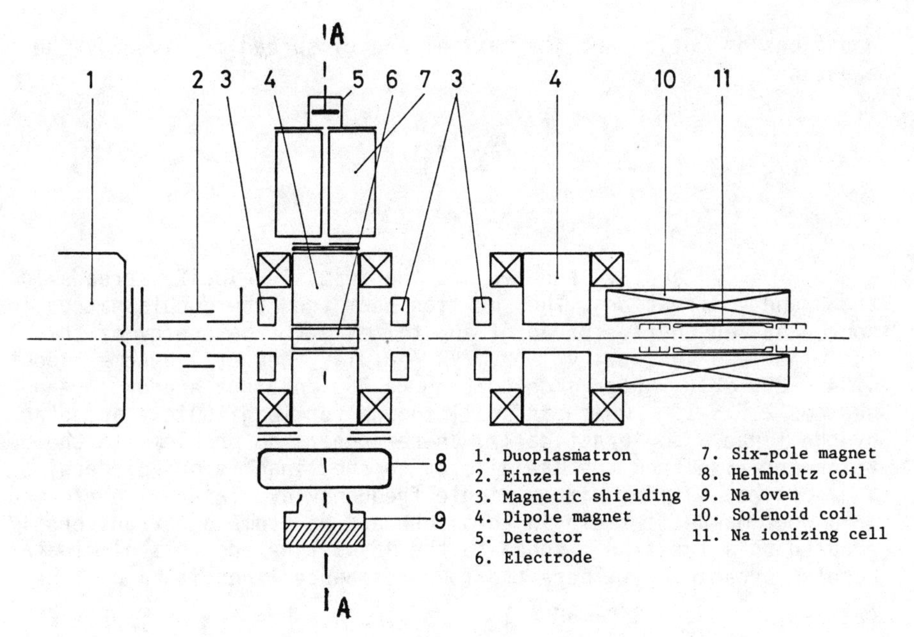

Fig. 9. Schematic layout of E × B scheme.

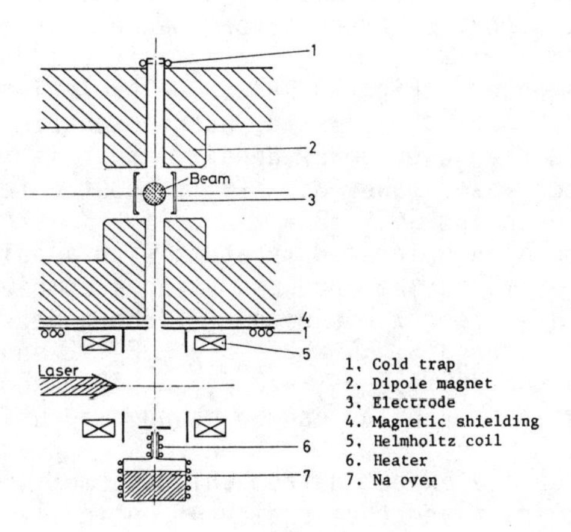

Fig. 9a. Cross sectional view of layout through A.

equations of motion and the maximum energy spread is given by the equation:

$$\Delta\varepsilon/\varepsilon = 2\left[\left(\frac{\Delta v_z}{v_0}\right)^2 + \left(\frac{\Delta v_y}{v_0}\right)^2\right]^{1/2},$$

where $v_0 = E/B$ and Δv_z and Δv_y are the initial velocity spreads in the z and y directions. The H^+ ion beam from the duoplasmatron ion source has an energy spread of about 1 eV. If the energy of the H^+ ion beam is 5 keV, $\Delta v_z/v_0 \sim \Delta v_y/v_0 \sim 0.014$, then $\Delta\varepsilon/\varepsilon$ becomes about 0.04 (~200 eV). After accelerating to 750 keV, the energy spread becomes 2.7×10^{-4} which is small enough for longitudinal acceptance by the linac. So, practically, there appears no problems if the beam transport line from the preinjector to the linac is non-dispersive.

Optical pumping with a single frequency dye laser is performed in a weak magnetic field in this scheme. Na atoms are transversely exposed to a laser beam tuned to the Na D1 line and this eliminates Doppler broadening. There are four resonance lines in Na D1 line:

(a) $3^2S_{1/2}F = 1 \rightarrow 3^2P_{1/2}F = 1$ (b) $3^2S_{1/2}F = 1 \rightarrow 3^2P_{1/2}F = 2$

(c) $3^2S_{1/2}F = 2 \rightarrow 3^2P_{1/2}F = 1$ (d) $3^2S_{1/2}F = 2 \rightarrow 3^2P_{1/2}F = 2$

In these resonances, a left (σ^+) or right (σ^-) circularly polarized laser beam changes the population distribution between the magnetic substates m_F. When the atomic states pass adiabatically from the weak magnetic field region to the strong magnetic field region, the atoms have then $m_J = +1/2$ or $-1/2$, where m_J is the z-component of atomic angular momentum. Fig. 10 shows the evolutions of the population of $m_J = +1/2$ or $-1/2$ atoms for each resonance as a function of $\beta \cdot t$, where β is the laser power density and t is the interaction time. The required laser power density is about a few mW/cm^2, and a σ^-laser beam tuned to the $3^2S_{1/2}F = 2 \rightarrow 3^2P_{1/2}\ F = 2$ resonance should make 100% electron-spin polarized Na atoms with a polarization of -1. The polarization of +1 can be produced using a combination of two single frequency dye lasers: one with σ^+ light tuned to the wave length of the resonance $3^2S_{1/2}F = 2 \rightarrow 3^2P_{1/2}F = 2$ and the other with σ^+ light tuned to the $3^2S_{1/2}F = 1 \rightarrow 3^2P_{1/2}F = 2$ resonance. In this way, the polarization direction can be changed by using two single frequency lasers.

The beam intensity expected from this scheme is determined by the H^+ ion beam current and the density of optically pumped Na atoms. The maximum density of Na atoms is limited by the imprisonment resonance and it is about 2×10^{12} cm^{-3} when the width of the Na

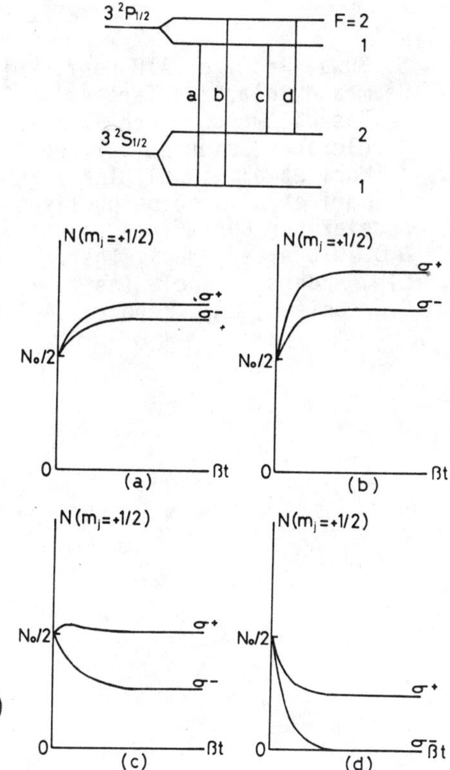

Fig. 10. Behavior of population of electron-spin polarized (+ 1/2) states as a function of βt.

atomic beam is 1 cm. If the length of the charge-exchange region is 5 cm, and the total conversion efficiency from H^+ ions to H^- ions is about 0.54% then 5.4 μA of polarized H^- ion current can be obtained if the H^+ ion beam current is 1 mA.

Practically, this scheme has many technical problems to be overcome. One is the problem of discharge in the E × B field region. Recently, we have tested the discharge of Na vapor in the E × B field region and found that discharge did not occur up to a Na atom density of 10^{13} cm^{-3} when the voltages were applied in a pulsed mode operation. The pulse duration and repetition rate were 250 μsec and 20 Hz. Another problem is the fringing field effect at the entrance of the E × B region. It is necessary to make a careful study of this effect. It is considered that many other technical problems must be overcome to realize this scheme. However, it seems to be a possible way to increase the proton-spin polarization. Another possible idea, which has been proposed at this workshop, is the scheme using an ECR ion source. In an ECR ion source, the H^+ ion beam can be extracted in a strong longitudinal magnetic field, so that the emittance blow-up problem can be eliminated. This scheme may be a good way to increase the proton-spin polarization and beam intensity simultaneously.

REFERENCES

1. S. Suwa, Proc. of AIP Conf. "High Energy Physics with Polarized Beam and Polarized Targets", 325, (1978).
2. H. Sasaki, Proc. of the U.S.-Japan Seminar on High Energy Accelerator Science, 247, (1973).
3. Y. Mori et al., Nucl. Inst. Meth., 141, 383, (1977).
4. Y. Mori et al., to be published in Proc. of 1981 Particle Accelerator Conference, Washington.
5. G.J. Witteveen, Nucl. Inst. Meth., 158, 57, (1979).
6. L.W. Anderson, Nucl. Inst. Meth., 167, 363, (1979).
7. G.G. Ohlsen, LASL report, LA-3949 (1968).

The CERN Polarized Atomic Hydrogen Beam Target

L. Dick, W. Kubischta, CERN

A polarized atomic hydrogen beam target has been developed at CERN to be used as an internal target for the CERN-SPS (Ref. 1).

A schematic diagram is shown in Fig. 1. A microwave dissociator is followed by a sextupole and an RF-transition 2-4. Contrary to a typical atomic beam design for a classical polarized ion source, a second sextupole eliminates state 4 and provides a better focussing over the rather short distance (~ 15 cm) to the beam dump. The main advantage of this unusual feature is the fact that the beam consisting almost exclusively of atoms in state 1 requires only a guide field of a few gauss. Therefore, the spin orientation can easily be switched to any direction, and the useful aperture is not restricted by large coils.

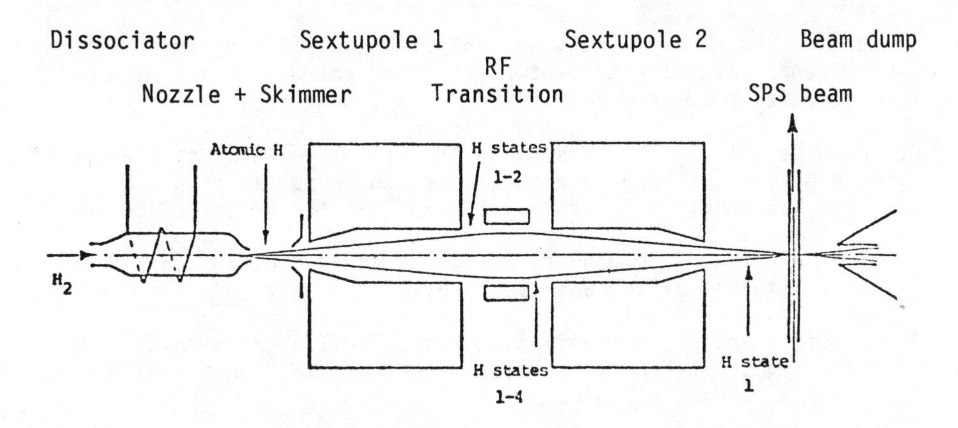

Fig. 1. Schematic diagram of the polarized hydrogen jet target

The target layout has been designed for function in the SPS-tunnel, taking into account the requirements of the vacuum environment of the accelerator (a more detailed description is given in Ref. 1).

Only a very short series of measurements were made, giving a density of about 10^{11} atoms/cm^3 (dissociator at room temperature), with a profile corresponding very well to the calculated distribution. The value above is substantially below the expected density of up to 10^{12} atoms/cm^3, but due to a change in program of the group no further work could be done on this polarized gas target.

While now being mainly involved in the preparation of experiments using an unpolarized gas target (Ref. 2), we have started a small development program with the aim of producing a very cold atomic beam. The interest in this subject was triggered by the publication by several groups of results on storage of very cold hydrogen atoms (Ref. 3,4). Even if the extreme temperatures reached in these experiments do not seem essential for our application, we can make use of other experimental information, especially on recombination on low temperature walls. These aspects have been broadly treated during this workshop by D. Kleppner.

We are, at present, setting up a test bench to explore the basic problems of the production of a very low temperature atomic beam, such as dissociator efficiency and atomic velocity. A second step would then be the choice of a suitable beam optical system (solenoid, sextupole or other).

REFERENCES

1. L. Dick, J.B. Jeanneret, W. Kubischta, The CERN Polarized Atomic Hydrogen Target, Proc. Int. Symp. High Energy Physics with Polarized Beams and Polarized Targets, Lausanne, 1980.

2. CERN-Lausanne-Michigan-Rockefeller Collaboration, Proposal for the e^+e^-, γ, π° and hyperon production in $\overline{p}p$ reactions at $\sqrt{s}$ = 22.5 GeV using an internal jet target at the SPS, SPSC/80-63.

3. T.J. Greytak, D. Kleppner, Bull. Am. Phys. Soc. 23, 86 (1978). L.F. Silvera, J.T.M. Walraven, Phys. Rev. Lett. 44, 164 (1980).

4. L. Dick, private communication, cited in T.O. Niinikoski, Proc. Int. Symp. High Energy Physics with Polarized Beams and Polarized Targets, Lausanne, 1980.

Notes Concerning Polarized Ion Sources*

G.J. Witteveen
Holec Gas Generators, B.V., Holland

Atomic beam source:

- Cooling of the dissociator will increase the ion beam intensity with $T^{-1/2}$ (thesis, page 12), when the multipole magnet is properly designed.
  This dependence is confirmed by other and our experiments.

- Magnetic field strength in classical electro-magnets is restricted to about 1.2T for 6-pole magnets.
  We reached this value in our Laboratory.
  Use of a compressor magnet is more important when cooling down the dissociator. I might point out the possibility of improving the "compression working" by superimposing a 4-pole field up on the 6-pole field, resulting in a negative, spherical aberration.

- The steep increase of the ionization efficiency (thesis, page 18, fig. 5a) can be fully explained by the calculations of Smith and Hartman (thesis, page 21, ref. 15).

  In accordance to these calculations and the performance of crossed E×B vacuum meters, we estimate an optimal ionization efficiency at electron entrance energy of 1.5 -2.5 KeV and a magnetic field between 0.5 and 1 T. The emittance degrading effect of this large magnetic field can be partly compensated:
  a) Take a magnetic field which is increasing towards the end of the ionizer as it should be for proper ion extraction.
     The protons are now focussed in a cross-over just behind the place where the electrons are reflected.
  b) Neutralize the protons in a short sodium target at optimum energy in a constant magnetic field with a direction reversed to that in the ionizer.
  c) The second electron is transferred in another charge-exchanger with a proper magnetic gradient. Doing so there will be, by proper electric focussing, a correlation between the magnetic field at which the proton was created in the ionizer, the distance travelled to the "second" charge exchange and the magnetic field at which this second charge exchange occurs.
- Using a very low cooled dissociator nozzle and a rather high magnetic field, it is possible to increase the energy of the magnetic dipole above the thermal energy, i.e. $\mu_B \cdot H > kT$.
  In this case electron-spin-up atoms are reflected by a strong axial or transversal field at the exit of the dissociator. In the case of an axial magnetic field it might be tried by properly shaping of the magnetic field to focus one spin state immediately into the ionization region.

*From a letter to A.D. Krisch dated May 14, 1981.

Crossed beam source:

- The polarized beam intensity we obtained was about 75% of the intensity which might be expected at most. This means no difficulties should arise when making an estimation of what is possible with this kind of polarized ion source.
  The beam emittance will be close to the emittance of the primary proton source when ion optics, magnetic fields and distances between the two charge exchange regions are carefully matched.
- The Achilles' heel of this kind of source of course is the obtainable density of polarized sodium atoms. When I proposed to construct this kind of polarized ion source in 1971, the first design was foreseen to be equipped with a 5W dye-laser. Unfortunately this laser was too expensive for us.
  However, about a year ago, just before finishing my work on polarized ion sources, we reached a sodium density of about $8 \cdot 10^{11}$ (at/cm^3) with a 1.2 T 6-pole magnet. Using a laser might increase this density about 5 times by decrease of the distance between the oven and the exchange region. A more important possibility might be an injection of the polarized sodium beam into a long and narrow charge exchange cell if the polarization is not lost quickly by wall collisions. I discussed this possibility some years ago with Dr. Steffens from Heidelberg.
- As to our polarization measurements we found about 1/3 of the maximum value. The magnetic field strength in the first charge-exchange region was about 0.1 T.
  Although we did not know exactly the distribution of the states which are created on neutralization of the protons, I still believe that a large loss of polarization is due to the bad vacuum, at least $4 \cdot 10^{-6}$ torr, in the charge exchange cell.

Thesis: "Some developments in polarized ion sources" by G.J. Witteveen, Technische Hogeschool Eindhoven, 1979.

AIP Conference Proceedings

		L.C. Number	ISBN
No.1	Feedback and Dynamic Control of Plasmas	70-141596	0-88318-100-2
No.2	Particles and Fields - 1971 (Rochester)	71-184662	0-88318-101-0
No.3	Thermal Expansion - 1971 (Corning)	72-76970	0-88318-102-9
No.4	Superconductivity in d-and f-Band Metals (Rochester, 1971)	74-18879	0-88318-103-7
No.5	Magnetism and Magnetic Materials - 1971 (2 parts) (Chicago)	59-2468	0-88318-104-5
No.6	Particle Physics (Irvine, 1971)	72-81239	0-88318-105-3
No.7	Exploring the History of Nuclear Physics	72-81883	0-88318-106-1
No.8	Experimental Meson Spectroscopy - 1972	72-88226	0-88318-107-X
No.9	Cyclotrons - 1972 (Vancouver)	72-92798	0-88318-108-8
No.10	Magnetism and Magnetic Materials - 1972	72-623469	0-88318-109-6
No.11	Transport Phenomena - 1973 (Brown University Conference)	73-80682	0-88318-110-X
No.12	Experiments on High Energy Particle Collisions - 1973 (Vanderbilt Conference)	73-81705	0-88318-111-8
No.13	π-π Scattering - 1973 (Tallahassee Conference)	73-81704	0-88318-112-6
No.14	Particles and Fields - 1973 (APS/DPF Berkeley)	73-91923	0-88318-113-4
No.15	High Energy Collisions - 1973 (Stony Brook)	73-92324	0-88318-114-2
No.16	Causality and Physical Theories (Wayne State University, 1973)	73-93420	0-88318-115-0
No.17	Thermal Expansion - 1973 (lake of the Ozarks)	73-94415	0-88318-116-9
No.18	Magnetism and Magnetic Materials - 1973 (2 parts) (Boston)	59-2468	0-88318-117-7
No.19	Physics and the Energy Problem - 1974 (APS Chicago)	73-94416	0-88318-118-5
No.20	Tetrahedrally Bonded Amorphous Semiconductors (Yorktown Heights, 1974)	74-80145	0-88318-119-3
No.21	Experimental Meson Spectroscopy - 1974 (Boston)	74-82628	0-88318-120-7
No.22	Neutrinos - 1974 (Philadelphia)	74-82413	0-88318-121-5
No.23	Particles and Fields - 1974 (APS/DPF Williamsburg)	74-27575	0-88318-122-3
No.24	Magnetism and Magnetic Materials - 1974 (20th Annual Conference, San Francisco)	75-2647	0-88318-123-1
No.25	Efficient Use of Energy (The APS Studies on the Technical Aspects of the More Efficient Use of Energy)	75-18227	0-88318-124-X

No.26	High-Energy Physics and Nuclear Structure - 1975 (Santa Fe and Los Alamos)	75-26411	0-88318-125-8
No.27	Topics in Statistical Mechanics and Biophysics: A Memorial to Julius L. Jackson (Wayne State University, 1975)	75-36309	0-88318-126-6
No.28	Physics and Our World: A Symposium in Honor of Victor F. Weisskopf (M.I.T., 1974)	76-7207	0-88318-127-4
No.29	Magnetism and Magnetic Materials - 1975 (21st Annual Conference, Philadelphia)	76-10931	0-88318-128-2
No.30	Particle Searches and Discoveries - 1976 (Vanderbilt Conference)	76-19949	0-88318-129-0
No.31	Structure and Excitations of Amorphous Solids (Williamsburg, VA., 1976)	76-22279	0-88318-130-4
No.32	Materials Technology - 1976 (APS New York Meeting)	76-27967	0-88318-131-2
No.33	Meson-Nuclear Physics - 1976 (Carnegie-Mellon Conference)	76-26811	0-88318-132-0
No.34	Magnetism and Magnetic Materials - 1976 (Joint MMM-Intermag Conference, Pittsburgh)	76-47106	0-88318-133-9
No.35	High Energy Physics with Polarized Beams and Targets (Argonne, 1976)	76-50181	0-88318-134-7
No.36	Momentum Wave Functions - 1976 (Indiana University)	77-82145	0-88318-135-5
No.37	Weak Interaction Physics - 1977 (Indiana University)	77-83344	0-88318-136-3
No.38	Workshop on New Directions in Mossbauer Spectroscopy (Argonne, 1977)	77-90635	0-88318-137-1
No.39	Physics Careers, Employment and Education (Penn State, 1977)	77-94053	0-88318-138-X
No.40	Electrical Transport and Optical Properties of Inhomogeneous Media (Ohio State University, 1977)	78-54319	0-88318-139-8
No.41	Nucleon-Nucleon Interactions - 1977 (Vancouver)	78-54249	0-88318-140-1
No.42	Higher Energy Polarized Proton Beams (Ann Arbor, 1977)	78-55682	0-88318-141-X
No.43	Particles and Fields - 1977 (APS/DPF, Argonne)	78-55683	0-88318-142-8
No.44	Future Trends in Superconductive Electronics (Charlottesville, 1978)	77-9240	0-88318-143-6
No.45	New Results in High Energy Physics - 1978 (Vanderbilt Conference)	78-67196	0-88318-144-4
No.46	Topics in Nonlinear Dynamics (La Jolla Institute)	78-057870	0-88318-145-2
No.47	Clustering Aspects of Nuclear Structure and Nuclear Reactions (Winnepeg, 1978)	78-64942	0-88318-146-0
No.48	Current Trends in the Theory of Fields (Tallahassee, 1978)	78-72948	0-88318-147-9
No.49	Cosmic Rays and Particle Physics - 1978 (Bartol Conference)	79-50489	0-88318-148-7

No. 75	Low Energy X-ray Diagnostics-1981 (Monterey)	81-69841	0-88318-174-6
No. 76	Nonlinear Properties of Internal Waves (La Jolla Institute, 1981)	81-71062	0-88318-175-4
No. 77	Gamma Ray Transients and Related Astrophysical Phenomena (La Jolla Institute, 1981)	81-71543	0-88318-176-2
No. 78	Shock Waves in Condensed Matter - 1981 (Menlo Park)	82-70014	0-88318-177-0
No. 79	Pion Production and Absorption in Nuclei - 1981 (Indiana University Cyclotron Facility)	82-70678	0-88318-178-9
No. 80	Polarized Proton Ion Sources (Ann Arbor, 1981)	82-71025	0-88318-179-7
No. 81	Particles and Fields - 1981: Testing the Standard Model (APS/DPF, Santa Cruz)		0-88318-180-0
No. 82	Interpretation of Climate and Photochemical Models, Ozone and Temperature Measurements (La Jolla Institute, 1981)		0-88318-181-9